The
Football

The Football

The Amazing Mathematics of the World's Most Watched Object

Étienne Ghys

Translated by Teresa Lavender Fagan

Princeton University Press

Princeton & Oxford

Original French edition LA PETITE HISTOIRE DU BALLON DE FOOT
(c) ODILE JACOB, 2023

English translation copyright (c) 2025 by Princeton University Press

Published by Princeton University Press
41 William Street, Princeton, New Jersey 08540
99 Banbury Road, Oxford OX2 6JX

press.princeton.edu

GPSR Authorized Representative: Easy Access System Europe -
Mustamäe tee 50, 10621 Tallinn, Estonia, gpsr.requests@easproject.com

Library of Congress Control Number: 2025936226

ISBN 9780691263120
ISBN (e-book) 9780691278933

British Library Cataloging-in-Publication Data is available

Editorial: Diana Gillooly, Whitney Rauenhorst
Production Editorial: Elizabeth Byrd
Cover: Ben Higgins
Production: Erin Suydam
Interior Design: Wanda España
Publicity: Matthew Taylor (US), Kate Farquhar-Thomson (UK)
Copyeditor: Rebecca Faith

Cover Credit: dpa picture alliance / Alamy Stock Photo;
textures from Adobe Stock

Printed in the United States of America

1 3 5 7 9 10 8 6 4 2

MIX
Paper | Supporting
responsible forestry
FSC   FSC® C008955
www.fsc.org

*For Guillaume, a football fan,
and perhaps a bit less a fan of geometry . . .*

*This little book may make him look at footballs
differently, and see the beauty in them!*

Contents

Preface

Many football enthusiasts never look closely at the ball: they're happy just to kick it. The classic ball has white and black panels. How many of each? Ask your friends (and yourself) that question, and you'll find that very few people know the answer. We're going to examine footballs from all the angles.

What is geometry for? It allows us to understand what is around us, beginning with very familiar objects, like footballs. Thanks to geometry, we can better situate ourselves in space, we can read a map, or find our way through a city.

We live among many geometrical objects, even if we don't always see them. The horizon is a straight line, walls are vertical, a water level is horizontal, doors and windows are rectangles. The sun and the full moon appear to be disks in the sky, but we know that's an illusion and that they are in fact spheres of which we can see only one surface. Balls are spheres and have been symbols of perfection for a very long time.

Studying geometry allows us to understand all these things. And to understand, we must observe and reflect. We must acclimate our brains to perceive every aspect of an object when it is displaced in space. We have to draw too, of course. After all, drawing and geometry have long gone hand in hand. In 1611, Lodovico Cigoli wrote to his friend Galileo:

A mathematician, however great,
Without drawing, will be not only half a mathematician,
But also a being without eyes.

This little book contains a lot of images. They are actually more important than the text.

An image is worth a thousand words.

That is especially true of geometry.

Étienne Ghys

Translator's Note

"Soccer" or "football"? The history of both terms is fascinating, and I encourage readers to delve more deeply into it. An illuminating explanation by John M. Cunningham called "Why Do Some People Call Football 'Soccer'?" can be found at Britannica.com. For the purposes of this delightful—and very informative—little book, and for the benefit of a majority of Anglophone readers and fans, I have chosen to use "football" throughout.

1

The Ball

Footballs often have black and white panels.

We're going to count them.

But why did people go to the trouble of making such complicated balls?

That is what we are going to try to understand in the following chapters.

We have to begin by taking the time to observe the ball in detail.

Even professional illustrators make mistakes.

THE TV STAR

Here's a photo of an old football from around fifty years ago. It is made of leather, shaped by panels sewn together. It was inspired by the ball from the 1970 World Cup in Mexico and has subsequently served as a model for almost all footballs.

Its name is *Telstar* because it was the star of television! During that period World Cup matches were televised for the first time. Perhaps it was also called *Telstar* because that is the name of one of the first telecommunication satellites, which was very famous in the 1960s.

The black panels have five sides, and the white panels have six. How many black panels are there? It's not easy to count them because you only see one side of the ball, and you have to imagine the other side. There is one black panel in the middle and five others around that one. So you see six black panels. How many are there on the other side, the side you don't see at all? Well, there must be the same number. Six panels are hidden. That makes a total of twelve.

The ball contains twelve black panels, each having five sides.

For the white panels, look again at the ball, but put a white panel in the middle.

The middle white panel touches three other white panels, and we can see six more near the edge, which means we see $1 + 3 + 6 = 10$. On the other side, there are ten more.

The ball has twenty white panels, each with six sides.

You might have some doubts about this, because there are white panels we can't completely see, which are a bit on each side. Maybe we counted them twice?

There are twelve black panels. Each one touches five white panels. Does that make $12 \times 5 = 60$ white panels? No, because each white panel touches three black panels. If we count that way, we come up with three times too many white panels. So, we have to divide those 60 by 3. That gives us twenty white panels. The count is good!

But the simplest and most accurate way to find out is to pick up an actual ball and count. Even in geometry, reasoning isn't always enough, and experience is important.

In total, our ball above is formed of $12 + 20 = 32$ leather panels that had to be sewn together. Let's count the number of sewn sides. The twelve black panels have $12 \times 5 = 60$ sides, and the twenty white panels have $20 \times 6 = 120$ sides, which comes to a total of 180. Since each seam connects two sides, you have to divide by 2.

That comes to a total of 90 seams.

SOME VERY ASTONISHING MISTAKES

It isn't easy to depict the *Telstar* ball. I asked some kids between the ages of 6 and 12 to draw a ball that was placed in front of them. Some drawings are pretty close, and others are . . . more imaginative.

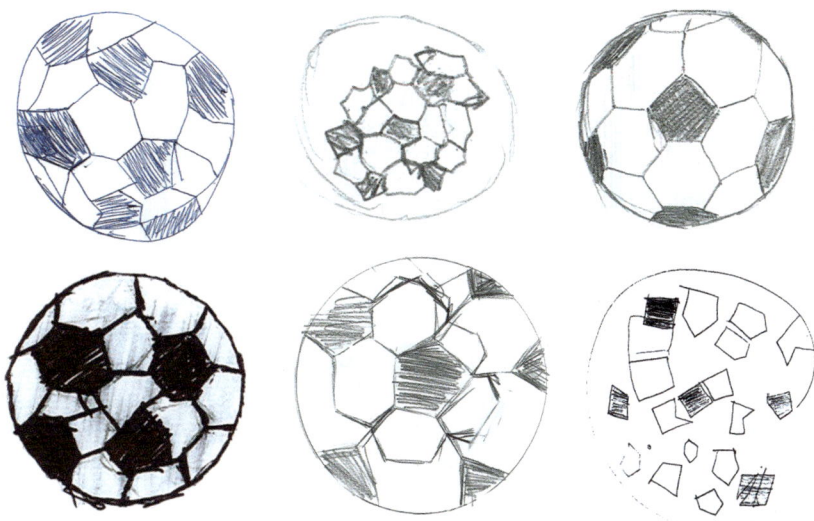

The "football" was invented in England over one hundred fifty years ago. One might therefore assume that the English know their footballs pretty well. And yet . . .

Look at the traffic sign posted in England indicating the way to a football stadium.

Unbelievable! "Shocking!" as the English might say. All the panels, the white and the black, have six sides.

That's impossible, that drawing is a mistake and doesn't depict a true football: the illustrator was wrong.

In 2017, mathematicians protested to the English government and asked that the sign be corrected.

Here's the petition that garnered 22,536 signatures:

> *The football shown on UK street signs (for football grounds) is made entirely of hexagons. But it is mathematically impossible to construct a ball using only hexagons. Changing this to the correct pattern of hexagons and pentagons would help raise public awareness and appreciation of geometry.*

Alas, the English government didn't see it that way. Here is its response:

> The Department for Transport commissioned research into road user's understanding of traffic signs in 2011. This concluded that respondents "showed a good basic level of understanding as to what different types of sign meant" and recommended that signs should be kept simple.
>
> The purpose of a traffic sign is not to raise public appreciation and awareness of geometry which is better dealt with in other ways. If the correct geometry were put onto a sign, it would only be visible close up and not from the distance at which drivers will see the sign. The detail of the geometry would also not be taken in by most drivers who were merely looking at the sign for direction. The higher level of attention needed to understand the geometry could distract a driver's view away from the road for longer than necessary which could therefore increase the risk of an incident.
>
> Additionally the public funding required to change every football sign nationally would place an unreasonable financial burden on local authorities. The Department could not justify the spending needed as an exercise to increase public awareness and appreciation of geometry.
>
> For the reasons given, we will not be changing the football symbol used on a traffic sign.
>
> —Department for Transport

Don't be misled by the drawing above which appears to show a football, all of whose panels have six sides. By attempting to complete the figure on the back side, you would encounter some problems.

The Champions League ball is also very important. It is attractive, with its twelve five-pointed stars and twenty white six-sided panels. Almost like *Telstar*.

But the Champions League logo isn't correct.

There are indeed five-pointed stars, but on the actual ball, there are three stars that surround a white six-sided panel. On the logo, some stars are correctly grouped three-by-three, but others are grouped four-by-four, which creates eight-sided panels, not the correct six. The design is wrong!

Even professional illustrators make mistakes when they depict a football, even when they have it right before their eyes (or better yet under their feet).

Here's the logo for Qatar's candidacy for the 2022 World Cup:

In the middle, all is well: a five-sided polygon surrounded by five other polygons, like a *Telstar* that one views with a black panel in the center. Around it, there are indeed the requisite thirty-two polygons, arranged in eight spirals, each containing four polygons. But all these polygons have five sides, and we know that there are only twelve of them that have five sides, whereas the others have six sides.

Clearly, no one really observes soccer balls, not even the people drawing them.

A SQUARE BALL?

Contemporary artists often use objects that are familiar and transform them into works of art. It is precisely because we don't really look at them carefully that artists want to show us just how significant they are for us.

Fabrice Hyber created hundreds of objects that he calls *Prototypes d'objets en fonctionnement* (Prototypes of working objects), or *POF*. His square football is POF 65, and I'm told the artist organizes actual football matches using such balls. Players probably have a lot of fun with them. After all, rugby is played with oval balls, so why not square balls?

However, there are two things I would like to point out to Fabrice Hyber.

First, his ball isn't square, it's cubic. You draw a square on a sheet of paper, but in space you talk about a cube.

Second, it would be closer to an actual football if the black panels had five sides, not six.

2 The Cube and the Five Elements

It seems pretty obvious that *Telstar*'s thirty-two polygons aren't assembled haphazardly.

There appears to be a sort of regularity: all the black panels are placed in the same way, surrounded by five white panels.

Why is that?

Could it be done differently?

Are other balls possible?

Mathematicians have been pondering those questions for a very long time, well before the invention of the football.

We are now going to create objects that resemble footballs, even if they are a bit too pointy. Later, we will see how to make real footballs by modifying them slightly.

These objects have bizarre names: tetrahedron, hexahedron, octahedron, dodecahedron, icosahedron. We'll even do a little "refresher" in Latin and Greek to understand them.

They are called the five "Platonic solids," in honor of one of the greatest thinkers of all time, who lived over two thousand years ago.

THE CUBE

Before we can understand the *Telstar* football, with its thirty-two panels or faces, we must begin by describing the familiar cube: it's easier, because a cube has only six faces, which are squares. You'll see in the end that it's not a silly way to start, because modern footballs are also cubes!

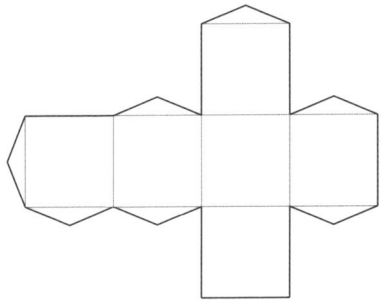

To make a cube out of paper, the way children do in elementary school, you cut out a cross shape made of six squares, as in the figure above, keeping in mind the little tabs on the sides. Then you fold and glue the tabs. Done!

The cross pattern is not the only possibility. There are eleven ways to arrange the squares to make a cube.

A cube has six faces that are squares. On the image below you can see three, and there are three others on the opposite side. There are eight vertices, indicated here by little balls (we can only see seven).

There are twelve edges, shown here by bars that connect the vertices and that are at the sides of the faces. Four of them are on the top face, four on the underside, and four others are vertical. In the image, three are hidden.

Of course, it is not yet a ball. But imagine if this cube were made out of rubber and were inflated, like a balloon. You'd get something like this:

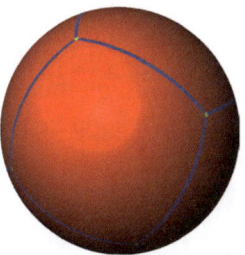

That looks more like a ball. But leather isn't rubber and even if we tried with rubber, we wouldn't get a ball as round as in the image above.

A LITTLE LATIN AND GREEK DICTIONARY

English is teeming with complicated words that have been passed down from antiquity, some from Greek, some from Latin. Often, Latin borrowed words from Greek, which further complicates things.

To set us on our way, first, here are five words that are useful to know in geometry.

GREEK	LATIN	ENGLISH
	latere	side
	aequi-	equal
-hedron		face
poly-		many
-gon		angle

And so, a polygon has many angles, and a polyhedron has many faces.

Let's move on to Greek and Latin words for numbers. You might not be familiar with all the geometry words in the right-hand column, but you'll soon be putting names to faces in the following pages.

NUMBER	LATIN	GREEK	GEOMETRIC EXAMPLE IN ENGLISH
1	uni-	mono-	unit
2	du-	di- bi-	bisector
3	tri-	tri-	triangle
4	quadri-	tetra-	quadrilateral tetrahedron
5	quinque-	penta-	pentagon
6	sexa-	hexa-	hexagon hexahedron
7	septi-	hepta-	heptagon
8	octo-	octo-	octahedron octagon
9	novem-	ennea-	enneagon
10	dec-	deca-	decagon
12	duodec-	dodeca-	dodecagon dodecahedron
20	viginti-	icosi-	icosahedron

Now you're ready to decode that a hexagon has six angles, a pentagon has five, and that a tetrahedron has four faces and an icosahedron has twenty. The *Telstar* is a polyhedron whose faces are pentagons and hexagons. The words are complicated, but you get used to them quickly.

Instead of saying "cube" like everyone else, now you can also say "hexahedron"!

And diverting from geometry for a moment: Why do the months of September, October, November, and December begin with Latin prefixes for 7, 8, 9, and 10, when they are the 9th, 10th, 11th, and 12th months of the year? In Roman antiquity, the year didn't begin on January 1; it began on March 1, two months later than today.

WHAT ABOUT TRIANGLES?

A cube has six square faces. What if we tried to glue together triangles rather than squares?

Here's an equilateral triangle.

Equilateral means the three sides are the same length (equi + lateral). Let's try to glue together equilateral triangles to make a solid, like we did with a cube. For example, using four triangles, we get this:

This object is called a tetrahedron ("four faces," in Greek). Around each vertex we see three triangles. To construct it we can cut and glue together this:

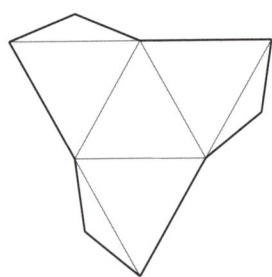

It has four faces, six edges and four vertices. It looks even less like a ball, even if it is inflated.

But you can also put four equilateral triangles around each vertex, like this:

This solid has eight triangular faces. It is an octahedron ("eight faces," in Greek). Here is the construction pattern:

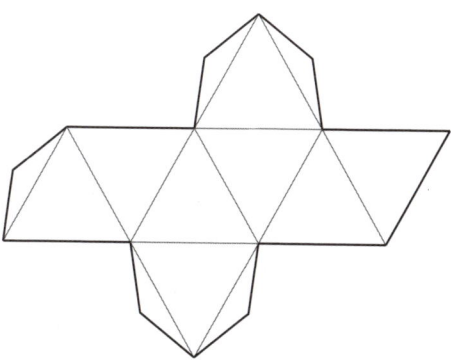

If we continue with five equilateral triangles around each vertex we get an icosahedron ("twenty faces," in Greek).

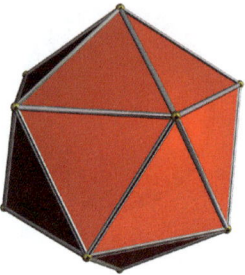

Twelve vertices, thirty edges, and twenty faces. Five triangles around each vertex.

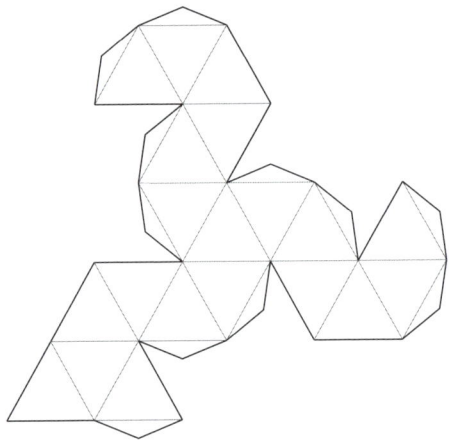

We have seen how to construct a cube, a tetrahedron, an octahedron, and an icosahedron. Up to now, it's the icosahedron that seems the roundest and which looks most like the *Telstar* football.

Maybe you've noticed that the icosahedron has twenty faces, the same number as the *Telstar*'s twenty white panels. This might not be a coincidence.

If we make it out of rubber and inflate it, since the faces are smaller, we might assume that it will be easier to obtain something round.

We've placed three triangles around one vertex, then four, then five. Can we continue with six equilateral triangles?

No, six won't work. If we place five equilateral triangles around a point, we can add a sixth, but all the space around the point will be filled, and we will no longer be able to fold the paper.

This is why with equilateral triangles we can only construct a tetrahedron, an octahedron, or an icosahedron, and nothing else!

AND WITH PENTAGONS?

We have used triangles (for tetrahedrons, octahedrons, and icosahedrons) and squares (for a cube). We can continue and try with pentagons. We can put three regular pentagons around a point, but no more than that.

By connecting the pentagons three-by-three around each vertex, we get a dodecahedron ("twelve faces," in Greek).

Here's the pattern to make it:

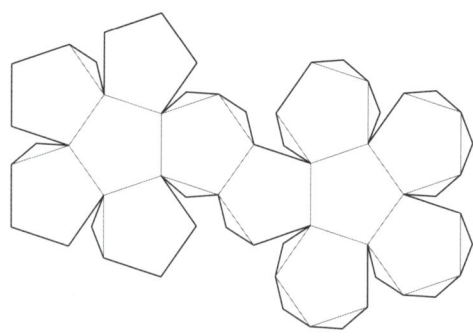

Here it is in inflated rubber:

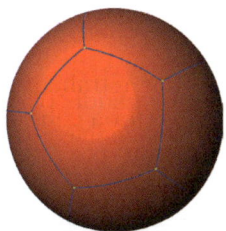

Twelve faces, thirty edges, and twenty vertices. Interesting! For the icosahedron, it was the opposite—twenty faces, thirty edges, and twelve vertices. Is this a coincidence?

Here is a drawing done by the great astronomer Johannes Kepler, precisely four hundred years ago.

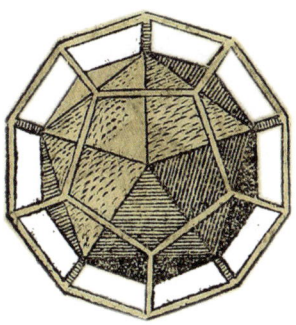

On the outside we see a dodecahedron with its twelve pentagons. Inside, there is an icosahedron with its twenty triangles. Each vertex of the icosahedron is in the center of a pentagonal face. Each vertex of the dodecahedron is above the center of a triangle. We can now understand why the number of faces of the one is equal to the number of vertices of the other.

In a certain sense we can say that the dodecahedron and the icosahedron are twins. In mathematics, we say that they are duals, or that one is the dual of the other.

This also works for the other polyhedrons that we have encountered.

If we put a vertex in the middle of each face of a cube, we get an octahedron. These two polyhedrons are thus duals.

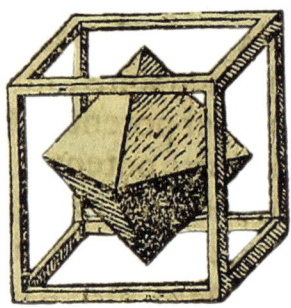

Cube: six faces, twelve edges, and eight vertices.

Octahedron: eight faces, twelve edges, and six vertices.

And what happens if we place a vertex in the middle of the faces of a tetrahedron? Well, we get another tetrahedron. The poor tetrahedron doesn't have a twin, it is the dual only of itself.

Tetrahedron: four faces, six edges, and four vertices.

Look at Kepler's drawing.

Johannes Kepler is known primarily as an astronomer who, at the beginning of the seventeenth century, discovered the laws that govern the movement of the planets. But his contributions go much farther, for example in optics, but also in our understanding of snowflakes, which physicists consider to be the foundational element in crystallography. His entire life was devoted to the concept of symmetry; in it he saw the key to understanding the universe.

Johannes Kepler (1571–1630)

A MAGIC TRICK

Make two rather thick cardboard models as shown below. Fold the sides of the center pentagon carefully.

Put one on top of the other and slip a rubber band around them, as indicated in blue on the photo on the right.

Using your finger, press on the middle to keep them together, as shown on the left. Let go quickly: a dodecahedron will appear before your eyes, like a rabbit jumping out of a magician's hat!

ICOSA- AND DODECAHEDRONS JUST ABOUT EVERYWHERE

Since the dodecahedron has twelve faces and there are twelve months in the year, we can make a calendar like this one:

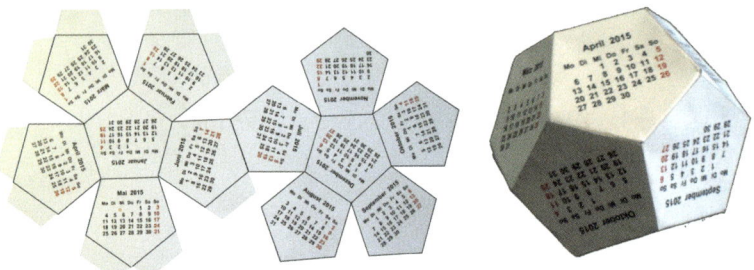

Here is a drawing of a radiolarian, a tiny living organism, which is found in the sea as zooplankton.

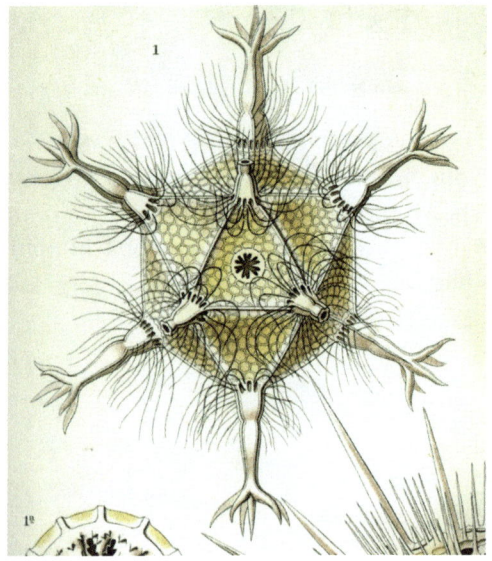

Many viruses, like this adenovirus, have the same shape.

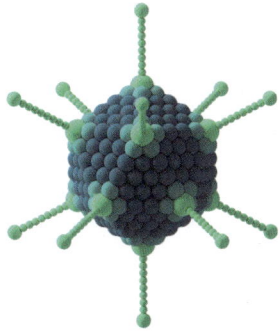

My glass dodecahedron reflects the little doily it is sitting on.

Now, here are three rectangles in space. Each has four vertices. In all, we have the twelve vertices of an icosahedron.

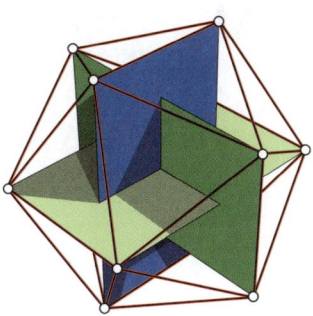

This form is even found in a monument dedicated to Spinoza in Amsterdam

and in this wooden model of a molecule.

THE FIVE ELEMENTS

 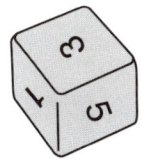   

We have glued together triangles, squares, and pentagons. Can we continue with polygons that have more sides?

No, that isn't possible.

If we try with regular hexagons, as soon as we put three together, we've filled the space around a point and we can't fold the paper anymore.

If we take regular polygons that have more than six sides, it's even worse, we can't even assemble three without overlapping them.

We now know that if we want to construct a regular polyhedron by assembling polygons that all have the same number of sides, there are only five possibilities. These polyhedrons are called "Platonic solids." Here they are, as drawn by Leonardo da Vinci five hundred years ago.

In the fourth century BCE, scholars wondered if one can cut matter infinitely small or if there are "grains" of elemental matter that cannot be further divided. The atomists—Leucippus, Democritus, Epicurus, Lucretius—maintained that everything surrounding us is formed by very small and indivisible particles (this is the etymological meaning of the word *atom*: "that which cannot be divided"). Their hypothesis was confirmed by modern science when we discovered and were able to see atoms.

However, there were other theories at the time. For other scholars, in particular Empedocles, the universe is made up of four substances or elements: earth, air, water, and fire, out of which everything else is composed. Plato even suggested a shape for each element.

- **Fire** pricks like a knife, thus it is a **tetrahedron**.

- **Air** is soft like an **octahedron**.

- **Earth** piles up easily like **cubes**.

- **Water** flows, like an icosahedron, the roundest of **polyhedrons**.

But what, then, is the role of the dodecahedron? It represents the entire universe. This was a very long time ago, and science was just getting under way.

Did Plato play football?

Plato (c. 428–348 BCE)

3 The Shape of the *Telstar* Ball . . . and Its Family

The Platonic solids are too pointy to make balls out of them. We have to cut off the points! We will now see how engineers created the magnificent *Telstar* ball of the 1970 World Cup.

Then, during the eight World Cups that followed, the ball didn't change shape very much. *Telstar* is indeed a star.

On the other hand, even when we cut off the points, our solids are still not very round, because the faces are flat, like the cardboard we used to construct them.

In real life, you inflate the ball so that the leather (or polyurethane) curves a bit.

TELSTAR

A football isn't a Platonic solid, because its faces aren't polygons that all have the same number of sides. It contains pentagons and hexagons. Also, it has to be inflated so that the faces aren't flat but rounded. You really do need a round football.

Let's start with the roundest Platonic solid: the icosahedron. Its problem is that it is too pointy with its twelve vertices. So, we cut off the points.

Since each vertex is surrounded by five triangles, we thus create twelve little pentagons that soften the points.

We've also cut off the three vertices of each of the twenty triangles, and the triangles then become hexagons whose sides are not all the same length: three sides are small and three are long.

If we cut off the points more and more deeply, the pentagons gradually grow until the moment when the six sides of the hexagons are the same length.

And voilà! *Telstar* is born.

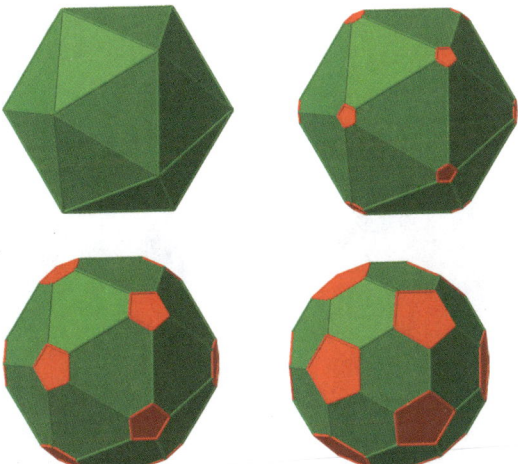

Twenty hexagons come from the twenty triangles of the ico-sahedron. Twelve pentagons come from the twelve vertices.

Another way of saying that we have cut off the points is to say that we have truncated the vertices. If we want to impress people with our knowledge, we can say:

Telstar *is a truncated icosahedron!*

Here is the pattern for cutting out the polygons:

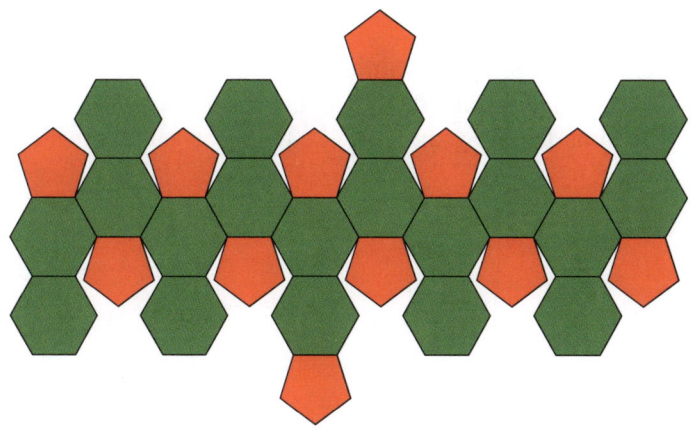

THE SYMMETRIES OF *TELSTAR*

Pick up a *Telstar* football.

First, place your two index fingers on two black pentagons on either side of the ball. Now, pivot the ball one-fifth of a turn without moving your fingers. The ball assumes an identical position: the white panels and the black panels have turned, but we have maintained the same position.

Do the same thing by placing your index fingers on two white hexagons. This time, do a one-third turn; the ball assumes the same position.

These are the ball's symmetries. Without this type of symmetry, the ball would fly off in bizarre directions when a striker aims for the goal, a bit like what can happen with an American football.

There is another interesting symmetry. Take a *Telstar* ball without any writing on it, like the one above.

Look at it in a mirror. What do you see? Exactly the same thing. This is another symmetry. My two hands are not identical. In a mirror, you seem to be left-handed if you are right-handed and right-handed if you are left-handed, but the same isn't true for *Telstar*: It is identical for left-handed and right-handed people.

WORLD CUP BALLS BETWEEN 1970 AND 2002

Footballs barely changed for thirty-two years.

1970, Mexico: *Telstar*

We know this ball well, it is the first football with thirty-two panels. It was a revolution in the world of football.

1998, France: *Tricolore*

In blue, white, and red! And one, and two, and three!
    The French know the rest.
    Vive la France!

TWO *TELSTAR*S: ONE OLD, ONE RECENT

People didn't have to wait for the invention of the football to create truncated icosahedrons. Here is a five-hundred-year-old drawing by Leonardo da Vinci, an illustration in a book by Luca Pacioli with a very lovely title: *Divine Proportion*. It doesn't discuss the symmetries of the football but rather the symmetries of the entire universe.

A much more recent discovery is a molecule formed by sixty carbon atoms. In French it is sometimes called the *foot-ballène* (and in English, the buckyball). Chemists discovered it about thirty-five years ago.

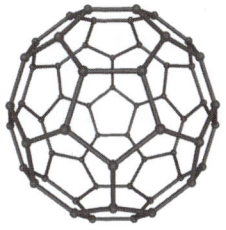

Leonardo da Vinci (1452–1519)

4 Footballs New and Old

Telstar is too complicated with its thirty-two panels.

In 2006, the engineers who design footballs wanted to simplify it. Or did they just feel it was time for a change?

They used solids discovered by Archimedes, another great thinker from antiquity. We'll now make their acquaintance.

My favorite football is the one from the 2014 World Cup in Brazil: it's called *Brazuca*. It is simply a cube and it is very elegant.

The first footballs before *Telstar* were also cubes that were greatly inflated to make them round. Isn't that hard to believe—cube-shaped footballs?

ARCHIMEDES

Archimedes was one of the greatest mathematicians of all time. He lived in the third century BCE. He of course was familiar with the five Platonic solids, whose faces are all polygons with the same number of sides, as we have seen. He looked for polyhedrons whose faces are regular polygons but not necessarily with the same number of sides. For example, the *Telstar*, a.k.a. "truncated icosahedron," has some hexagonal faces and others that are pentagonal.

Archimedes (287–212 BCE)

Archimedes found thirteen such examples that today are called "Archimedean solids." Here they are.

The drawing of the solid is on the left, and the pattern showing how to cut the paper to assemble it is on the right.

First, we can truncate the Platonic solids, as we did to make *Telstar*.

1. This one is made by cutting off the points of a tetrahedron. It is a truncated tetrahedron. It would serve as inspiration for the ball used in the 2010 World Cup in South Africa.

2. A truncated cube.

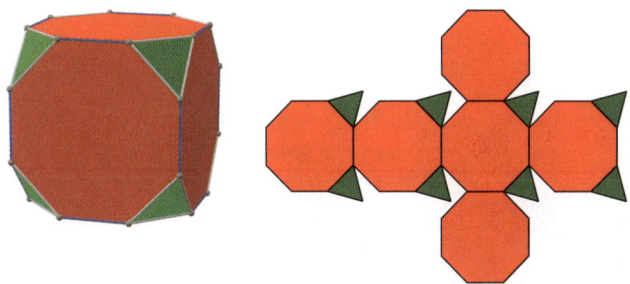

3. A truncated octahedron. We'll see it again when we describe the ball designed for the 2006 World Cup in Germany.

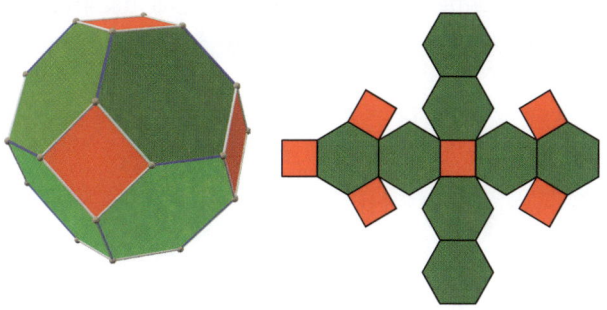

4. Here is our friend, *Telstar*! The truncated icosahedron.

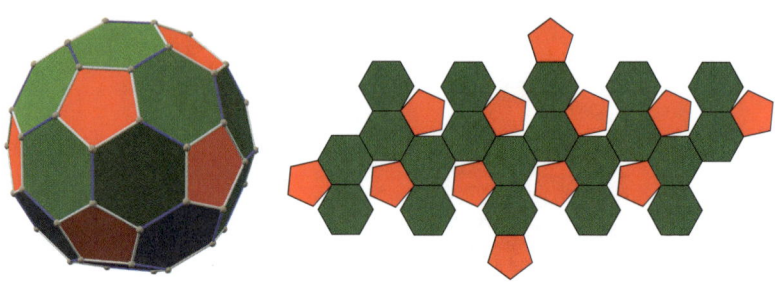

5. A truncated dodecahedron.

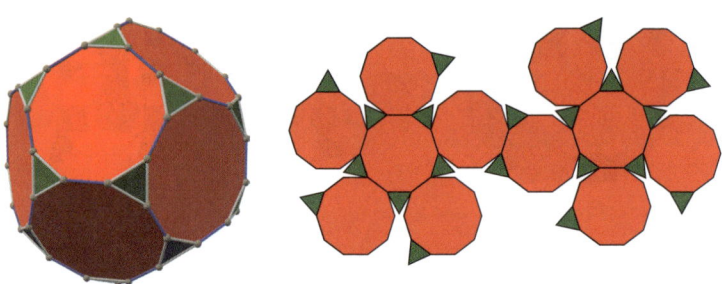

Then, the following solids have such complicated names that I prefer not to burden you with them!

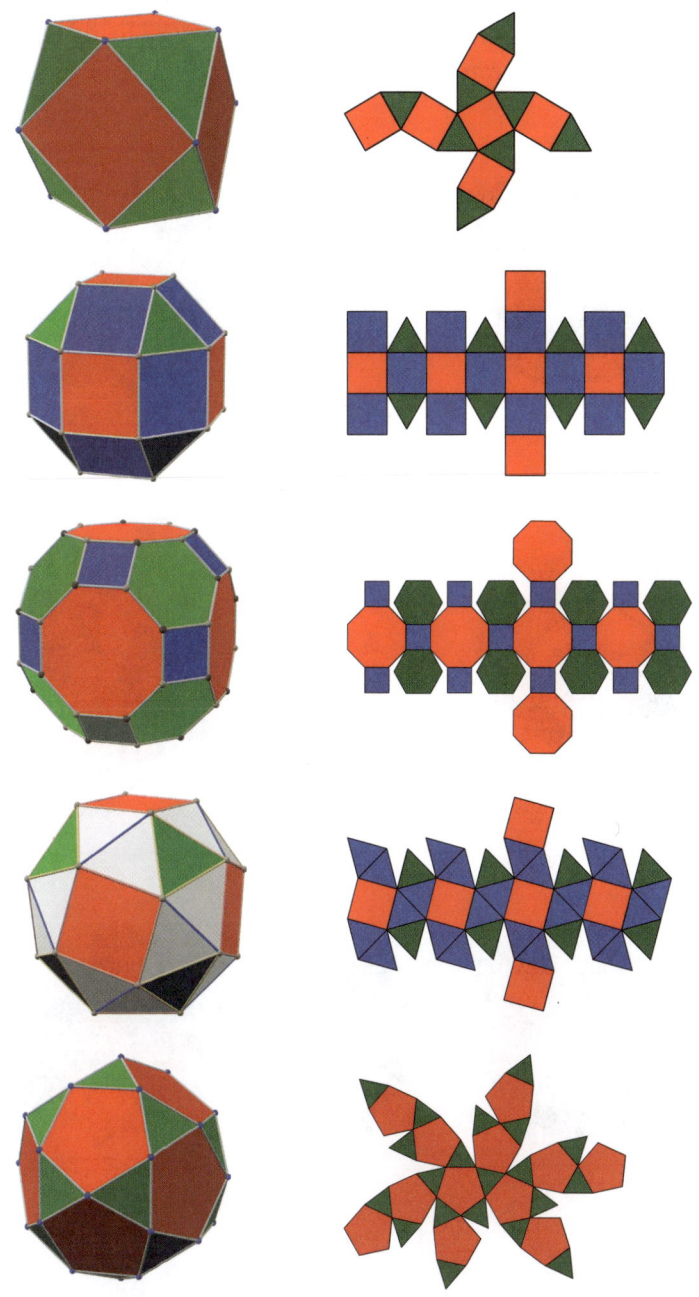

The ball of the World Cup in Qatar was inspired by this icosi-dodecahedron:

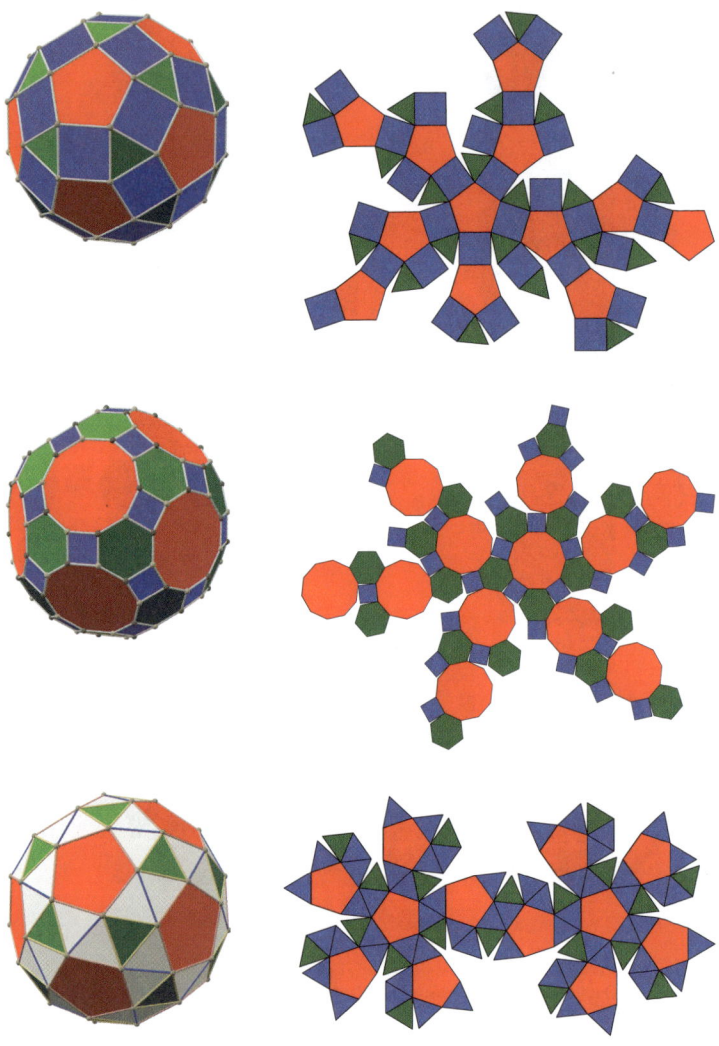

THE *TEAMGEIST* BALL FROM THE 2006 WORLD CUP IN GERMANY

By 2006, football makers were probably tired of *Telstar* and its thirty-two leather panels. They wanted to innovate by using fewer panels. They preferred to truncate a Platonic solid with fewer faces than the icosahedron.

They chose the octahedron, which has only eight faces and six vertices.

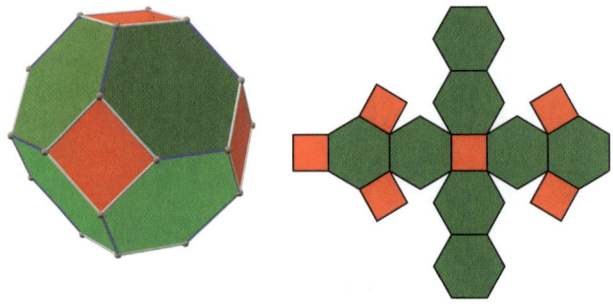

Once truncated, there are fourteen panels—eight hexagons and six squares.

The engineers had an excellent idea. Why should they cut out panels of polygons whose sides are straight lines. Why not cut out more curved shapes so that what they constructed would be round? And here is what they ended up with: eight hexagonal panels, three of whose six sides are curved. The six "square" panels are a bit bizarre, because they look more like beans.

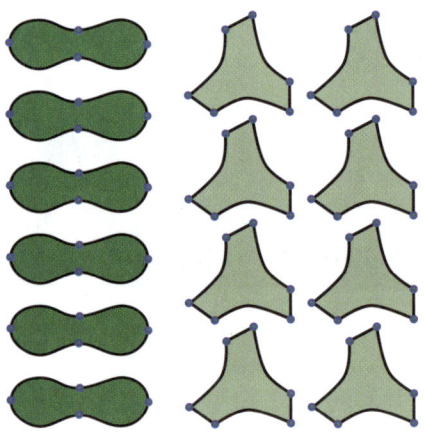

The six "squares" are placed like the vertices of an octahedron and the eight hexagons as the faces of an octahedron.

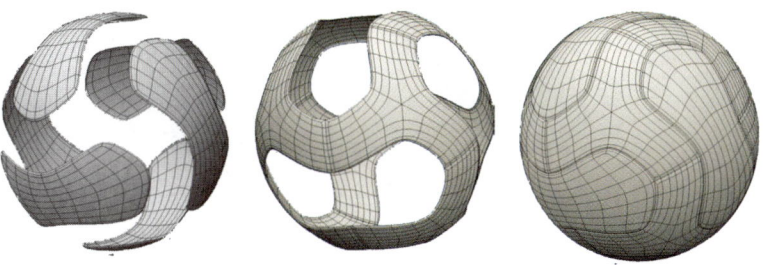

And here is everything together:

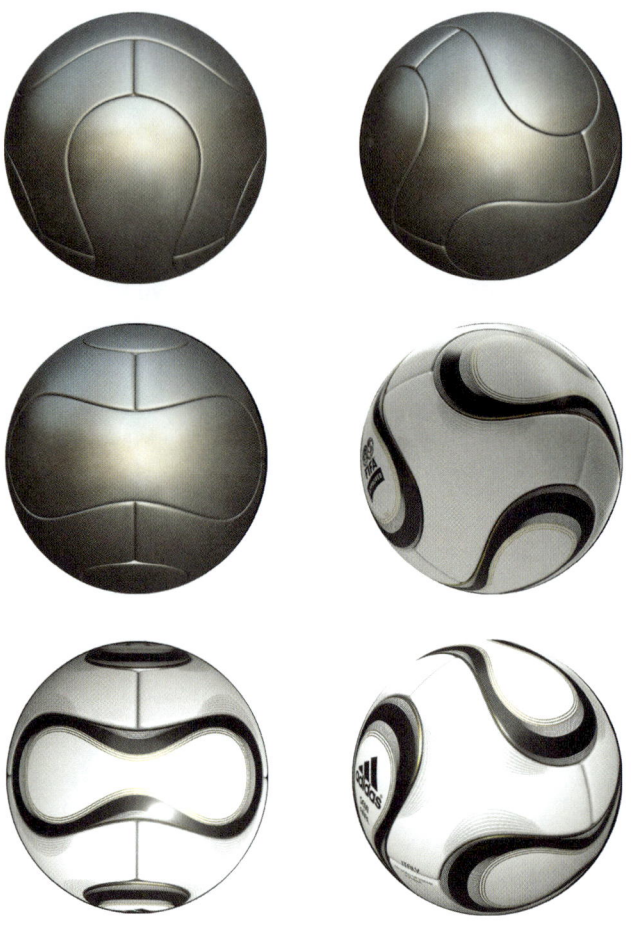

This ball is quite attractive, isn't it?

It's called *Teamgeist*, which in German means "team spirit."

But, as always, it's easy to criticize! Because *Teamgeist* has fewer seams than *Telstar*, it glides better in the air. Brazilian Roberto Carlos found that *Teamgeist* was too light and didn't perform well when it rained.

THE *JABULANI* BALL FROM THE 2010 WORLD CUP IN SOUTH AFRICA

This time, we begin with the tetrahedron, which has only four faces and four vertices. Once truncated, it has eight faces. But we have to admit that it isn't very round yet.

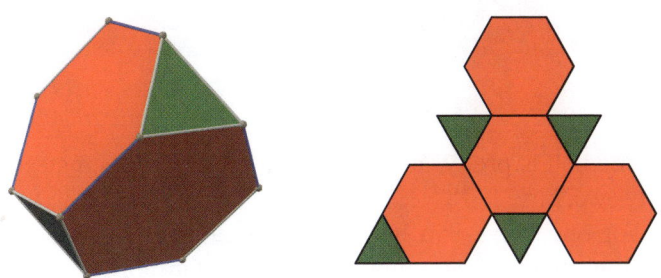

Designers developed nice shapes, as in the drawing below. There are eight panels, as for a truncated tetrahedron, but four are oval and four are hexagonal, with sides of different lengths.

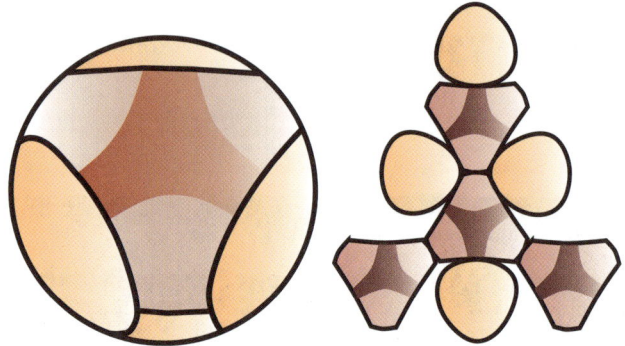

The result is the ball called *Jabulani*, which is a Zulu word meaning "rejoice." The version of the ball in the figure below is called *Jo'bulani*, a blend of "Jabulani" and "Jo'burg," a common nickname for Johannesburg, the site of the World Cup final.

The ball is pretty, and it uses only eight panels instead of thirty-two, but . . .

For one thing, you have to make panels that aren't flat, and that's not easy. For another thing, and far more important, players didn't like this ball. They found that it followed unpredictable trajectories. The French goalkeeper Hugo Lloris bluntly declared that the ball was a catastrophe.

It was a good idea to modernize *Telstar*, which was already thirty-six years old, but this wasn't a success. Ball designers continued on their quest.

THE *BRAZUCA* BALL FROM THE 2014 WORLD CUP IN BRAZIL

Telstar is a truncated icosahedron, *Jabulani* is a truncated tetrahedron, and *Teamgeist* is a truncated octahedron.

The designers decided to stop cutting off points and to return to the simplest polyhedron, the one that everyone knows: the cube.

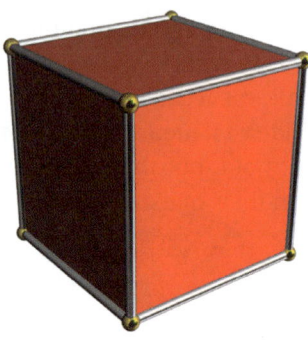

So, they had to find six "square" shapes that could be assembled like a cube, with the goal of forming a round ball. In other words, they were looking for a round cube!

Here are the squares they decided to assemble. Granted, they don't look like normal squares, but they have four vertices (in yellow) and, between the vertices, instead of having the usual straight sides (blue dotted lines), there is a nice, sinuous curve (red dotted lines).

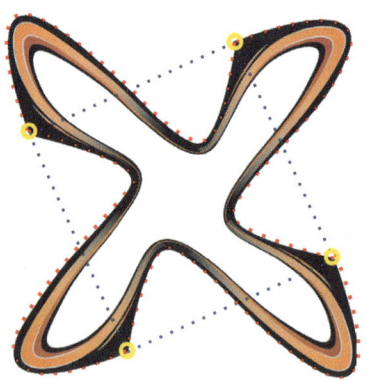

Then they assembled these "curved squares" exactly as one would do for a cube.

And they ended up with the *Brazuca* ball for the World Cup in Brazil. It is my favorite.

In the center of the image you can indeed see a "corner of a cube" where three faces meet. But that corner doesn't stick out in a point, because our squares are "curved."

So, if I say that this ball is a cube, I hope you'll believe me now!

The name of the ball, *Brazuca*, was chosen following a vote by more than a million Brazilians.

Two years later, the ball from Euro 2016 was still a cube but slightly different.

BALLS BEFORE 1970

In fact, almost all the footballs before 1970 were leather cubes, all of whose six faces were cut up into two or three rectangles.

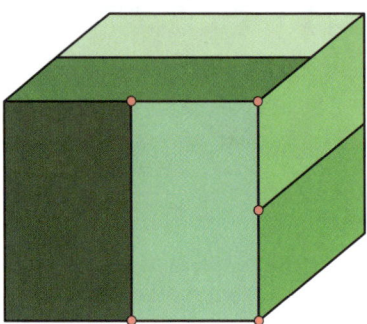

Of course, they had to really inflate the ball for it to be round.

In 1938, five of the faces of a cube were divided into two rectangles each, but one of the faces was divided into three triangles, in order to insert a valve to inflate the ball.

In 1962: squares and hexagons. What polyhedron is this?

A STRANGE CUBE

Let's go back to our cube whose faces have been divided into two rectangles. There are twelve rectangles, two on each of the cube's six faces. One might say that each rectangle has, in fact, five vertices. Of course, a rectangle has four vertices, not five, but look at this figure:

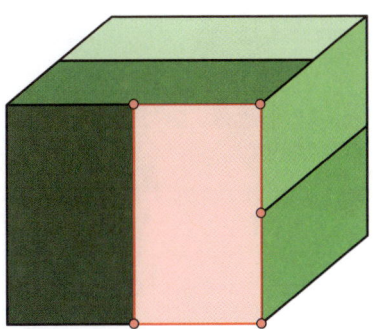

One of the sides of each rectangle has a sort of additional vertex in its middle. Thus, you have twelve "pentagons." If we count the vertices, we get twenty because there are the eight vertices of the cube and a new vertex in the middle of each of the twelve edges. At each of these twenty vertices, we see three "pentagons" that join up.

But we've already seen this! It is exactly like a dodecahedron.

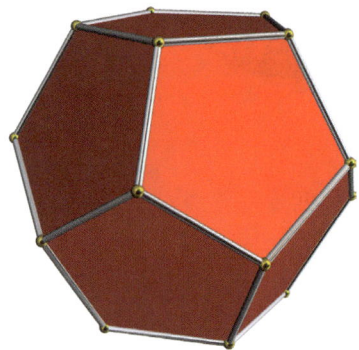

Our cube is not a dodecahedron, but they are constructed in the same way, from twelve panels that are regular pentagons for the icosahedron, and "pentagons in the form of rectangles" for the cube. The construction is the same: three panels around each of the twenty vertices.

Here is how we can deform our bizarre cube into a dodecahedron.

A very long time ago, Kepler noticed that if you place a roof on each of the six faces of a cube, you get a dodecahedron. Look at his drawing: in dotted lines you see a cube, and above you see the roof.

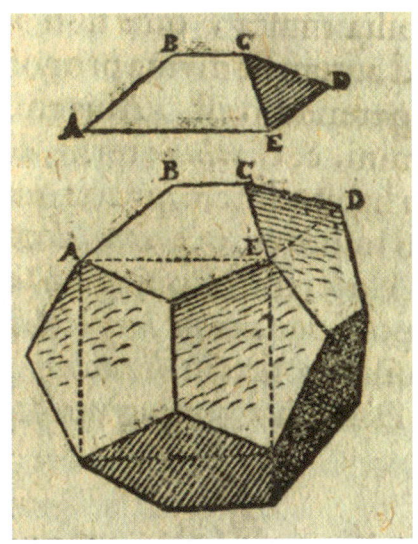

5 The 2022 World Cup Ball

Its name is *Al Rihla*, which means "the voyage" in Arabic.

To tell the truth, I'm a bit disappointed. Granted, *Al Rihla* is very attractive, but it is somewhat reminiscent of the past, almost identical to the *Telstar* of 1970, as we'll see.

It will provide an opportunity to meet a new polyhedron with a truly complicated name: the icosidodecahedron, which we have, by the way, already encountered as one of the Archimedean solids.

The ball's innovation thus goes back two thousand two hundred years! Archimedes would probably have been proud of it.

AL RIHLA: AN ICOSIDODECAHEDRON OR A CUBE?

Gone are the convoluted shapes we encountered with *Teamgeist, Jabulani,* and *Brazuca. Al Rihla* is constructed of eight triangular panels and twelve kite-shaped panels.

You may recall that *Telstar* balls are created from an icosa-hedron whose twelve vertices have been truncated. After the truncation, the vertices become regular pentagons, and the triangular faces (in green below) become hexagons (in blue). If we truncate even more deeply, cutting up to the middle of the edges, the hexagons become twenty equilateral triangles (in yellow).

You get an icosidodecahedron! *Icosi*- is Greek for "twenty," *dodeca*- for "twelve," and -*hedron* for "face." We've already seen it—it's one of the Archimedean solids.

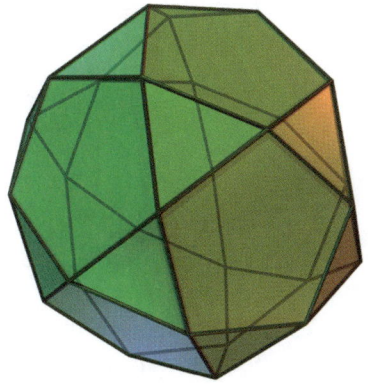

When it's inflated, you get this:

But we haven't yet achieved *Al Rihla*. Each of the red faces is surrounded by five green triangles. If we paint twelve well-chosen triangles red, we will have twelve red panels in the shape of kites and eight equilateral triangles will remain.

And you get *Al Rihla*. One might think that this ball is derived from the icosahedron, but actually we have broken the symmetries, as physicists would say: only eight equilateral triangles at the vertices of a cube remain. Here we are back at the cube. Remember Kepler's lovely drawing?

If you put a roof on the faces of a cube, you get a dodecahedron, which is the dual of the icosahedron. A large family.

Here is another view, in which we can imagine a cube.

So, how are the actual footballs made? Well, by putting together eight triangles and twelve slightly different-looking kites.

But that's not all. The people who designed this ball wanted to break the symmetries a bit more.

Compare a photo of the ball (on the left) and what we have just discussed (on the right).

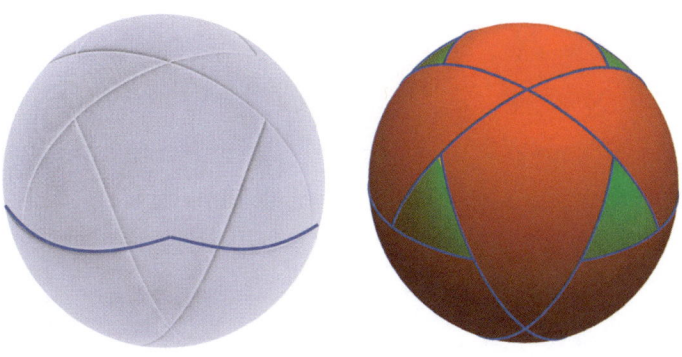

The difference is clear. In the image on the right, the bottom sides of the two triangles are on the same large circle (drawn in blue). On the left side, when you extend the bottoms of the two triangles, an angle appears. I prefer the right side, but they didn't ask for my opinion.

Upon reflection, this ball isn't very innovative. Look at this older ball. It was already an icosidodecahedron.

THE ELECTRONIC BALL

Here is another question. We have seen that the eight triangles form the vertices of a cube. Does this mean that the ball has the same symmetries as the cube? Here are two views in which a green triangle has been placed in the center.

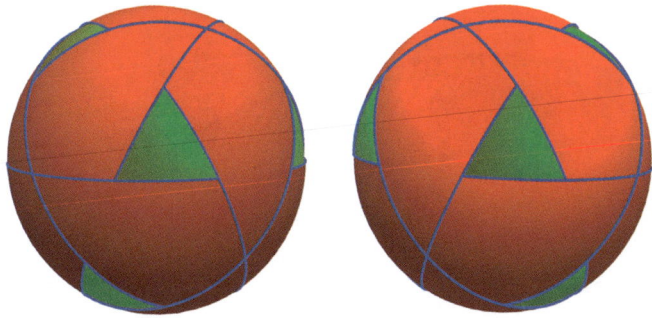

The kites don't turn in the same direction. There are two types of triangles. Again, it appears symmetry has been lost.

A final original feature of this ball steers us away from geometry. It is the first electronically connected ball! From inside the ball, a little device a few centimeters in diameter is connected by wires to the ball's surface. This device records all the ball's movements and can transmit them electronically to a computer that precisely records the ball's position at every moment. The referee can thus be informed objectively of offside plays.

6

A Bit of Physics

Up to now, we have only examined the geometry of footballs but not their actual physical properties, such as weight or dimensions. We have only been discussing abstract geometry.

It's now time to do some physics. We need to say a few, very incomplete, words about how balls fly through the air. The subject is complex, because understanding the flow of air around a ball in motion requires both experimental and theoretical study.

OFFICIAL DIMENSIONS OF "REAL" FOOTBALLS

Let's begin with dimensions. The official rules state that all balls must

- be spherical (this isn't American football).
- be made of suitable material (this isn't clear).
- be of a circumference of not more than 70 cm (28 in.) and not less than 68 cm (27 in.) (or a diameter of between 21.6 and 22.3 cm).
- be not more than 450 g (16 oz) in weight and not less than 410 g (14 oz) at the start of the match (which allows for a certain margin).
- be of a pressure equal to 0.6–1.1 atmosphere (600–1,100 g/cm²) at sea level (8.5 lb/sq in to 15.6 lb/sq in). (The ball's pressure cannot be lower than 1 atmosphere, otherwise the ball would implode. One must understand that the internal air pressure exceeds the external pressure by a value between 0.6 and 1.1 atmosphere, thus an internal pressure between 1.6 and 2.1 atmosphere.)

The rules for football are intentionally flexible regarding dimensions. This is particularly true for the field whose width can vary between 45 and 90 meters (49 and 98 yards) and length between 90 and 120 meters (98 and 131 yards). For international competitions, the boundaries are more precise: width between 64 and 75 meters (70 and 82 yards) and length between 100 and 110 meters (109 and 120 yards).

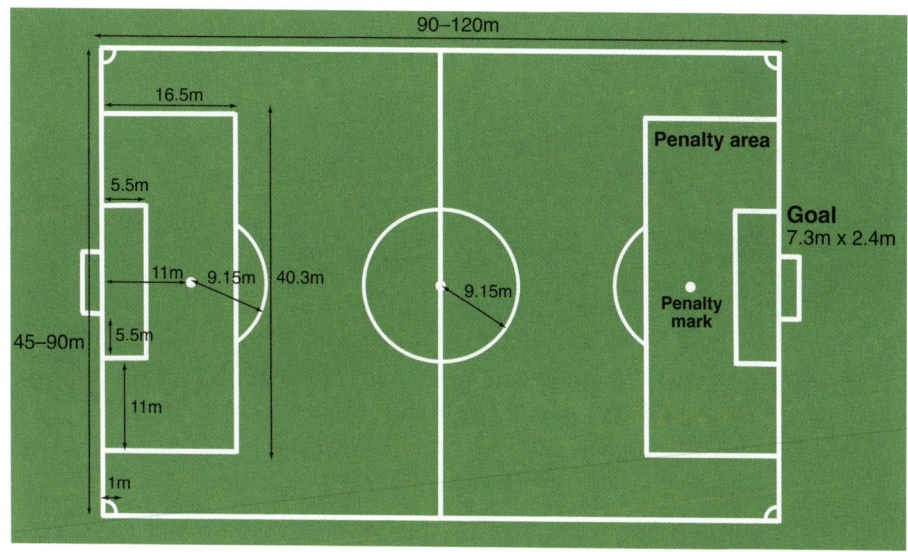

LITTLE RHOMBICOSIDODECAHEDRONS

Let's return one more time to the star of 1970: *Telstar*, with its regular hexagons and pentagons, created by truncating an icosahedron.

To be officially FIFA approved, the highest standard of quality, the ball must be spherical to within 1.5%. All points on a sphere are at the same distance from the center. To say that a ball is spherical to within 1.5% means that the difference between the maximum and the minimum distances from the center to points on the ball do not exceed 1.5% of the minimum distance.

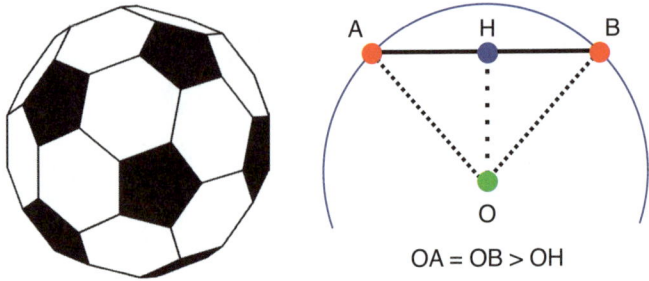

$$OA = OB > OH$$

The figure above clearly shows that the centers of the faces are closer to the ball's center than are the vertices. If we do the calculation, we find that the pentagons are 6.1% closer to the center than are the vertices, and the hexagons are 8.5% closer. Our truncated icosahedron is not spherical enough: FIFA will reject it. Of course, when you inflate the ball, the shape becomes more spherical, but perhaps not enough. Another problem is that the total surface occupied by the hexagons is 50% greater than that occupied by the pentagons.

So maybe the points of the icosahedron need to be truncated even further so that all the faces are at the same distance from the center. If we do that, our hexagons will no longer be regular, as in the following images. We can in fact see that the sides of the hexagons are of two different lengths.

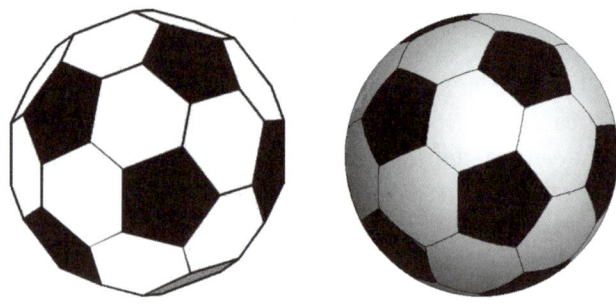

This ball shape is called isodistant. The Nike company registered a patent for this ball, but it seems their designers were mistaken in their calculations, which do not exactly present an equidistant model.

The ball can be made even rounder with more faces. Below is one of the Archimedean solids that we have already encountered but without its name: it is the little rhombicosidodecahedron. Here it is drawn by Augustin Hirschvogel for his 1543 book, *Geometria.*

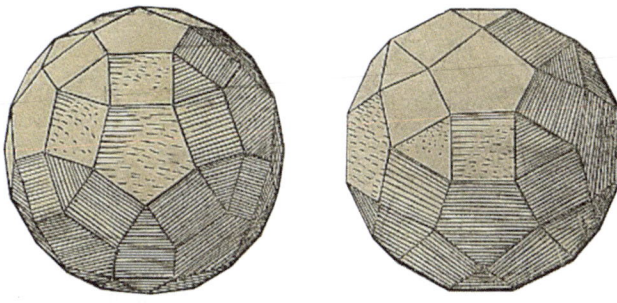

Its faces are not all at the same distance from the center, but if we adjust it adequately, as in the next image, we get a rounder ball.

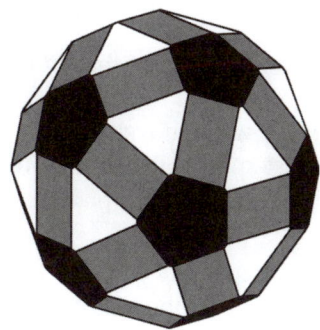

The geometric imagination of designers is limitless. Might they be copying Archimedes?

Surprisingly, the rules of football say nothing about the "suitable material" from which balls should be manufactured or about their texture. For a long time, the question wasn't asked. *Telstar* was made out of thirty-two leather panels that were sewn together. How would a smoother ball, made from a different material and with fewer seams, behave?

WHAT DO THE PLAYERS HAVE TO SAY?

Football players aren't shy about commenting on the qualities of any technological innovations when a new ball appears. Every time the ball changes, it takes time for players to get used to it.

Regarding the *Fevernova* ball of the 2002 World Cup in South Korea and Japan:

- Italian goalkeeper Gianluigi Buffon called it "a ridiculous bouncing ball for children."

- Brazilian Rivaldo declared that "it traveled too far."

In 2006, *Teamgeist* was a novelty, with only fourteen panels. This was the most spherical and smoothest ball ever conceived, and it was completely waterproof:

- German goalkeeper Oliver Kahn thought the ball was "constructed for the benefit of strikers."

- Roberto Carlos declared: "It is very light; the way it is made is completely different from before. It seems it is made out of plastic."

- According to David Beckham: "With this ball, you can make perfect crosses. And, for me, a free-kick specialist, that means I can roll the ball around the wall with an incredible precision."

In 2010, players were displeased with *Jabulani* and were unanimously against it! It was constructed of eight thermally bonded panels, each having a microtexture.

- Brazilian goalkeeper Júlio César: "It's terrible, horrible, it's like one of those balls you buy at the supermarket."

- Danish player Daniel Agger: "It makes us look like drunken sailors."

- Brazilian striker Robinho: "For sure, the guy who designed this ball never played football."

- Marcus Hahnemann: "You never know what's going to happen with this ball, it's everywhere."

- Luis Fabiano, the Brazilian striker: "All of a sudden, it changes trajectory on you. It's as if it didn't want to be kicked. It's incredible, it's as if someone were guiding it. You're about to kick it, and it moves aside. I think it's supernatural."

- Bert van Marwijk, Dutch manager: "The ball does the strangest things, especially when it moves through the air."

TARTAGLIA AND GALILEO

Well before the invention of football, people were interested in the trajectories of cannonballs. Science and war often go hand in hand.

Niccolò Fontana, a.k.a. Tartaglia, was an astonishing early sixteenth-century Italian scholar. It is possible that he discovered the solution to cubic equations, out of which he was later finagled by Gerolamo Cardano. But this dispute between mathematicians is diverting us from our subject.

In 1539, Tartaglia published a treatise on ballistics. Here is how he represents the trajectories of cannonballs launched in different directions. According to him, a cannonball initially goes in a straight line, before arcing toward the earth along a circular path, and finally falling in another vertical straight line directly toward the earth.

Niccolò Fontana, a.k.a. Tartaglia
(1499–1557)

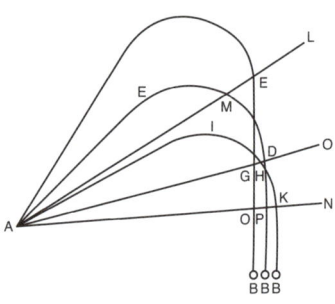

Today we know that that is not how a cannonball or a football behaves. Tartaglia didn't perform experiments; he reasoned using philosophical principles going back to antiquity. It's reminiscent of those cartoons in which a character runs off a cliff in a straight line with no longer anything under his feet; then he realizes with horror that he's in the void and ends up in a vertical free fall.

Galileo is one of the founders of modern science, in which experimentation and reasoning are combined. According to legend, he dropped cannonballs from the top of the Tower of Pisa to study falling objects.

A football in flight encounters several forces: its weight and the reaction forces of the air.

First, assume that those forces aren't present, which makes no sense, but will help us to understand. An object that is subjected to no force whatsoever moves at a constant speed in a fixed direction.

The trajectory is thus a straight line. Not very realistic.

Let's add weight, which is a constant force, directed downward. The weight of an object bends its straight trajectory downward: the ball follows a parabolic curve. Galileo is the one who figured that out.

Galileo (1564–1642)

In the figure below, we see the trajectories of a ball that goes at 10 m/s (or 36 km/h [22 mph]) in various directions. When it sets off vertically, it climbs to a height of 5 meters (16.5 ft). For it to go as far as possible, it must set off at a 45-degree angle, and it will travel up to 10 meters (33 ft).

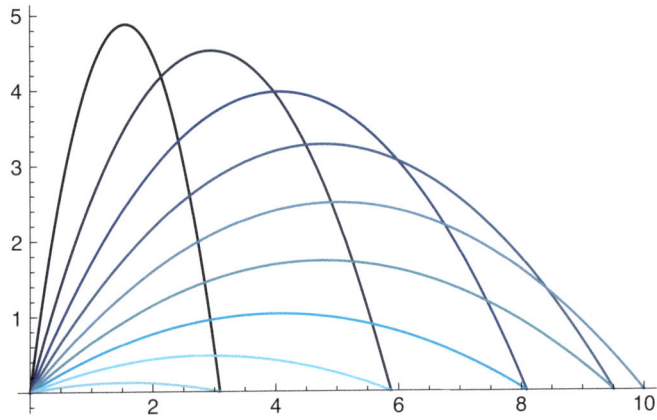

In fact, this parabolic curve is a pretty good description of the movement of a pétanque or bocce ball, because the air exerts very little resistance compared to the weight of the steel ball; but the trajectory of a football in the air is much more complex.

Let's add the friction of the air, which is called the drag. Now things get complicated. Drag is a force acting opposite to the relative motion of a moving object. We feel it when we put a hand out the window of a speeding car and a force pushes our hand backward. That is drag.

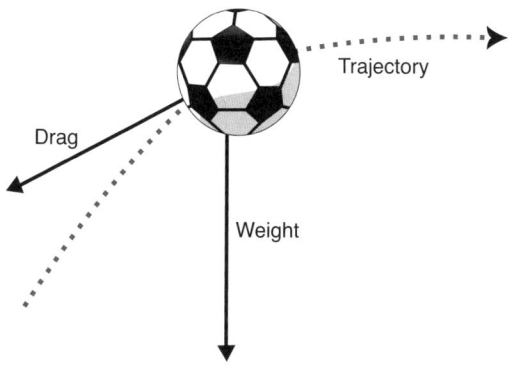

For a football moving at a speed of around 70 km/h (43.5 mph), drag and weight have approximately the same impact, unlike the pétanque or bocce ball whose weight dominates.

Consider the trajectory of a badminton shuttlecock going at 40 m/s (131 ft/s) at an angle of 60 degrees. When it hits the ground, its speed is only 6.7 m/s (22 ft/s). At the beginning, the curve resembles a parabola, but drag soon affects the trajectory, which looks more like a vertical fall at the end.

Ultimately, it looks more like the curves of Tartaglia.

BIRDS CAN'T FLY

The first theoretical studies on drag are quite old, going back at least as far as d'Alembert, who lived in the eighteenth century.

Jean le Rond d'Alembert is a central figure of the Enlightenment. In addition to his contributions in mathematics and physics, he was also one of the principal contributors, along with Diderot, to the important *Encyclopédie, ou dictionnaire raisonné des sciences, des arts, et des métiers*, which was published between 1751 and 1772.

Jean le Rond d'Alembert
(1717–1783)

Imagine a ball that is moving at a certain speed in the air. The air is disturbed by the ball and must be displaced to allow space for it. This involves forces of pressure all around the ball, the sum of which is the drag. In a famous article, d'Alembert mathematically calculated all these pressures and their sum. To his great surprise, the result of his calculation was ... 0! Thus, according to him, the air exerts *no* friction on a ball going through it. As we all know, that's impossible, a football slows down and there must be a force of friction.

This is d'Alembert's paradox:

A body that moves in the air encounters no frictional force.

This result is all the more surprising because it doesn't simply state that there is no drag, it also implies that there is no lift—the force that enables airplanes and birds to fly. Here, then, is the theorem of an illustrious mathematician that implies that birds cannot fly. How is this possible? Of course, d'Alembert had seen birds. At the end of his article, he explains that his theorem is correct mathematically, but that one must wait for other work to be done to understand why it doesn't apply to the real world.

Today, we would say that his reasoning was correct but the modeling of the interaction between the ball and the air does not correspond to physical reality—a nice example of the difference between physics and mathematics. Moreover, d'Alembert was fully aware of this and he "offered his paradox to geometricians."

XXXIVME MÉMOIRE.

Suite des Recherches fur le mouvement des Fluides.

§. I.

Paradoxe propofé aux Géometres fur la Réfiftance des Fluides.

1. SUPPOSONS un corps compofé de quatre parties égales & femblables, placé au milieu d'un fluide indéfini, lequel fluide foit renfermé dans un vafe recti-ligne. Imaginons que ce corps foit fixe & immobile ; & que les parties du fluide reçoivent toutes une impulfion égale, parallèle aux côtés du vafe & à l'axe du corps ; d'abord il eft évident que les particules du fluide à la partie antérieure, doivent .fe détourner & glifler le long du corps, & former des courbes d'autant plus approchantes de la ligne droite, qu'elles feront plus éloignées du corps, jufqu'à une certaine diftance (qui fera au moins celle des parois du vafe) où elles fe mouvront en ligne droite.

In d'Alembert's demonstration, there was a hidden hypothesis: it was implicitly assumed that air was a nonviscous fluid. We now understand that drag and lift are the results of viscosity, which is how much a fluid resists flowing.

Imagine a long tube through which air is blown at a certain speed. If air were nonviscous, it would continue its path at the same speed, without being aware, so to speak, of the walls of the tube, as shown in the following figure.

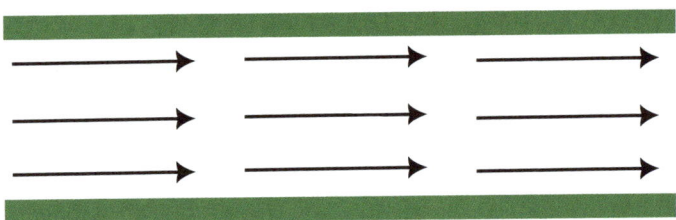

But viscosity causes air to stick to the walls and the speed of the air there is zero. The speed is highest in the middle and decreases gradually as it approaches the edge. If it isn't very obvious at first, think instead of the flow of honey in a tube.

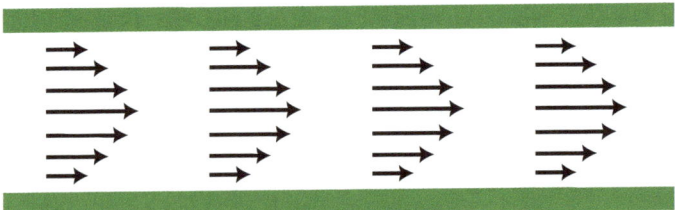

Viscosity can be surprising. For example, a glacier appears to be made of solid ice. However, if you took a photo once a year for twenty-five years and viewed the images one per second, you would see a stream flowing.

A famous (and long) experiment has been carried out in Australia since 1927. A funnel stopped up with a cork was filled with heated pitch, which is solid at room temperature. At first, they waited three years for the pitch to cool and settle, then they removed the cork. The pitch then flowed drop by drop, but one drop fell around every . . . eight years! Pitch is thus a highly viscous fluid, much more so than air. But it is in fact air's minute amount of viscosity that enables birds and planes to fly. D'Alembert's paradox isn't paradoxical after all.

GUSTAVE EIFFEL'S PARADOX

The first serious experiments studying the movement of a ball in the air were carried out by Gustave Eiffel at the beginning of the twentieth century, initially in a laboratory at the foot of his tower. The experiments involved observing objects that were dropped from the second level, as Galileo was believed to have done in Pisa (more likely in Padua, in fact, according to historians). The technological stakes were high because this was at the beginning of aviation and people were trying to understand how something "heavier than air" could effectively fly, contrary to what d'Alembert might have suggested. Of course, many of us see planes flying but don't really understand how it's possible!

Gustave Eiffel (1832–1923)

Later, Eiffel preferred to work in a different way. He understood that the forces that act on a ball moving in an immobile atmosphere are the same as those that act on a fixed ball in moving air. A physicist would call it Galileo's principle of relativity.

Instead of studying a ball moving through the air, it is much easier to take a fixed ball and place it in a wind tunnel. You can then measure the forces applied to the ball and photograph the trajectories of the air, for example, by introducing a bit of smoke.

Here is the layout of the experiment. A ball, hanging by two very thin strings, is placed in wind created by a fan whose speed can be adjusted. The force of drag exercised on the ball makes the strings lean backward at a certain angle. You can then measure the drag as a function of the wind speed but also as a function of the dimensions of the ball and the nature of its surface.

Fig. 58. — *Essais d'une sphère suspendue à des fils.*

When the wind speed is very low, viscosity is dominant. Imagine a marble that falls slowly in a jar of honey. The honey sticks to the entire surface of the marble and slows its fall. The same is true of little drops of rain that slowly fall from clouds, lingering on their viscous path. For a football flying through the air, this viscous interaction is produced only at extremely low speeds, at less than 1 km/h, which doesn't normally apply to a football.

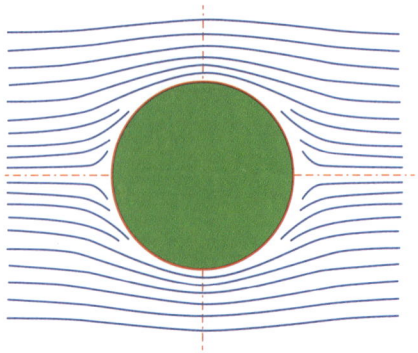

If the speed is higher, viscosity is no longer the only phenomenon that explains drag. Toward the front of the ball, under the effect of wind, it is as if a thin film of air, scarcely a few millimeters thick, were stuck to the surface like a second skin.

In this boundary layer, viscosity is dominant. On the internal surface of the layer, the air is in direct contact with the ball and sticks to the surface: it follows the ball's movement, at the same speed. On the external surface, viscosity doesn't really play a role—the air is too far from the ball, is disassociated from the surface, and the situation is approximately that of d'Alembert and his birds that can't fly. It is thanks to that thin layer that planes can fly.

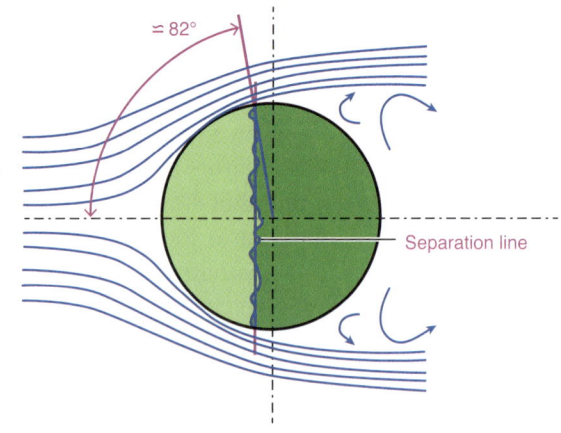

≅ 82°

Separation line

But this second skin that makes up the boundary layer doesn't surround the entire surface of the ball. In the back, the boundary layer no longer remains in contact with the ball: it is said to be released or separated. This occurs along a "separation line" located approximately halfway between the front and the back. That is when a wake develops—a zone in which not much really happens, apart from the fact that the pressure there is low. This resembles what we see on a river downstream from a bridge. The wake thus "breathes in" the ball and holds it. This is one of the origins of drag.

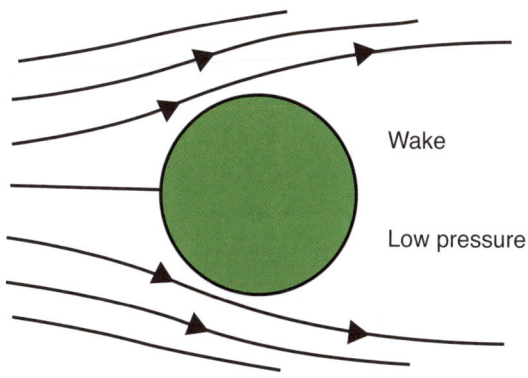

When the speed increases further, an interesting phenomenon occurs. Up to then, the circulation of the air was very regular; this is called laminar flow. But suddenly the air becomes turbulent inside the boundary layer, which means its movement appears irregular and erratic. The phenomenon is analogous to that of cigarette smoke that first climbs regularly before breaking and becoming turbulent.

The turbulent boundary layer adheres much better to the ball and the separation occurs closer to the back of the ball. The wake is smaller, it holds the ball less, and drag diminishes. The force holding the ball and slowing its movement decreases.

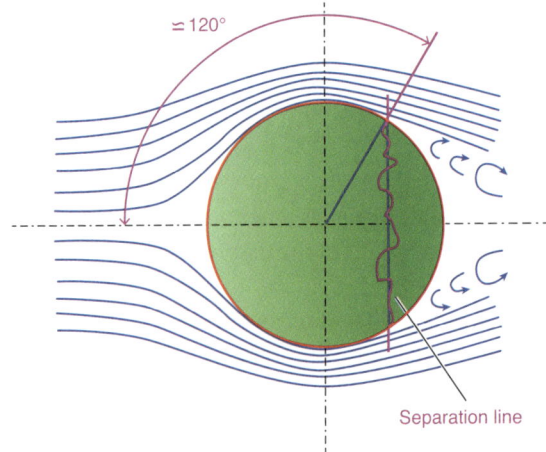

This phenomenon is clearly shown in the following photographs. The speed is greater in (b) than in (a), and the separation line is moved to the back.

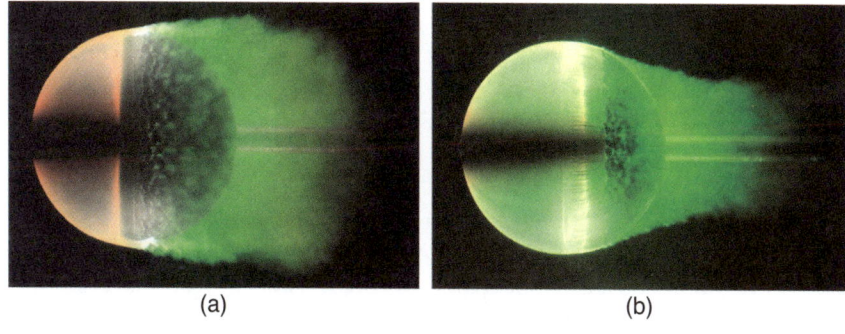

(a)   (b)

This is Eiffel's paradox, or the "crisis of drag."

When a ball exceeds a certain speed, its drag can decrease considerably.

In the beginning, Eiffel didn't believe this. To be convinced, he repeated his experiments many times. How could it be that by increasing speed the resistance of the air could diminish? For a football, the critical speed is around 10 or 12 m/s, or around 35 or 40 km/h (22 or 25 mph), a speed much lower than that at which a good player can make a penalty kick. When a player strikes a ball at high speed, or above the critical speed, the initial airflow is turbulent and drag is weak. The ball then slows only slightly, but when the speed decreases to the critical speed, the flow becomes laminar and drag suddenly increases. It can be multiplied by a factor of 4: the decrease in speed is then much greater. It gives the impression that the ball has stopped! In any case, the trajectory is broken. This is a phenomenon with which all players are familiar but that can adversely affect an inexperienced goalkeeper.

It is therefore important to know the critical speed of each ball.

THE ROUGH SURFACES OF FOOTBALLS

Wind tunnel experiments show that the critical moment for drag depends a great deal on how rough a ball's surface is. It so happens that an imaginary, completely smooth ball wouldn't make players happy: the air would tend to slow it down. The makers of the first leather balls were unaware of that fact.

Critical speed depends on

- the "suitable material" with which a ball is made, which can be more or less smooth;
- the total length of the panels' seams, and on their depth; and
- the shape of the panels.

All of that is experimental, and it seems impossible to develop a complete theory for it.

On recent balls, designers have introduced very small indentations in order to make the surface rougher, somewhat like the surfaces of golf balls. We can see this in the following photo of *Al Rihla*:

By the way, look at the rough surface of a golf ball. It looks like a bee hive, formed with hexagons. But you can also see some pentagons, if you look carefully.

Telstar has a total of more than 4 meters of seams, relatively well distributed over the entire ball, so that turbulence is produced at the same time over the entire surface. Its critical speed is around 12 meters/second (39 ft/s), well below that of a perfectly smooth ball of the same diameter, which is 37 meters/second (121 ft/s). Since most free kicks are made at speeds between 25 and 30 meters/second (82 and 98 ft/s), we can see that *Telstar* is less hindered than a smooth ball.

Teamgeist, on the other hand, has large smooth areas, and the passage to turbulence can be produced at different moments in different areas. As a result, the trajectory can become completely erratic and unpredictable.

Jabulani is the smoothest, with much shallower grooves than those of other balls. That's probably why the players didn't like it.

SPIN AND ROBERTO CARLOS

To understand the effect of spin, you must understand the Venturi effect. (Giovanni Battista Venturi was a late-eighteenth-century physicist.) Air is flowing through a tube of variable diameter. In the narrow portion, it accelerates because the amount of air flowing is constant; if there is less room for the same volume of air to flow, the flow must be more rapid. Pressure is measured on the walls in several places, and one notes that the pressure is lower where the tube is narrower. One then concludes that the pressure is lowest where the speed is greatest. This is the principle for measuring the speed of an airplane in flight. Air flows through in a small tube fixed on the fuselage and one need only measure the pressure to know the speed.

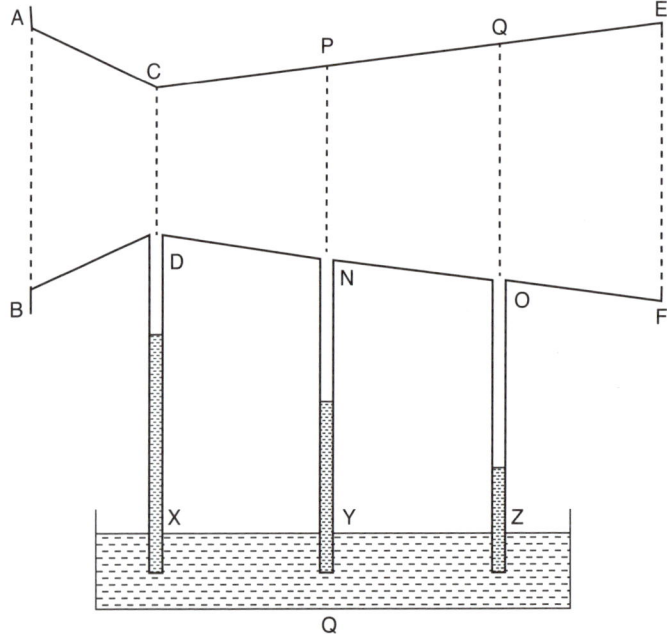

Now, imagine a ball spinning and flying at the same time (toward the goal?).

In the following figure, the ball is turning clockwise, so that on the upper part, thanks to viscosity, the rotation of the ball is combined with the speed of the wind. The opposite is true of the lower part. Pressure is therefore lower above, through the Venturi effect. The result is a force, lift, that is perpendicular to the speed of the ball, oriented upward in our case. This force is all the greater because the ball is turning quickly. The ball is in a certain sense drawn upward due to the lower pressure.

Of course, there is also drag directed toward the back, shown in green.

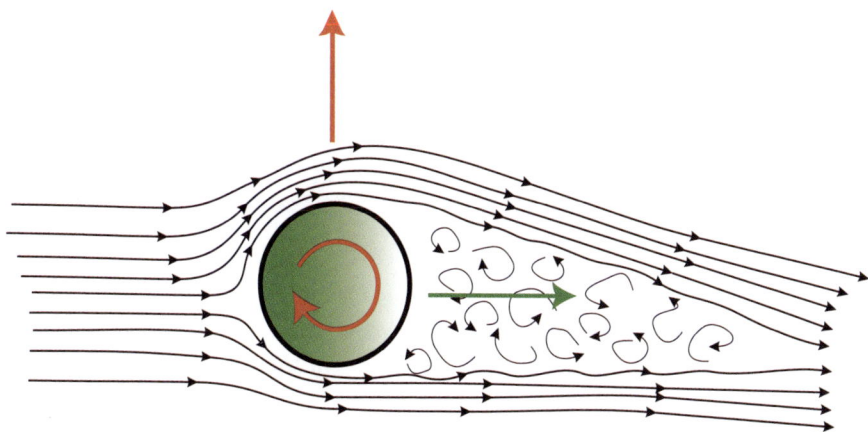

Everything depends on the ball's axis of rotation, which the experienced player can more or less choose when striking the ball: he can "brush" the ball with his foot to create rotation.

If the axis is horizontal, the lift force will not make the ball deviate to the side. The trajectory will remain in the same vertical plane and the effect of spin can then transform the trajectory into a spiral, even if in practice that seems very difficult.

If the axis is vertical, the trajectory can deviate from its initial plane, and this is how we can understand the historic goal of Roberto Carlos.

On Tuesday, June 3, 1997, at the Gerland stadium in Lyon, France hosted Brazil for the first match of the Tournoi de France, a warm-up to the 1998 World Cup. At 11:07 p.m., in the 22nd minute of play, Roberto Carlos had a free kick, 35 meters from the goal of Fabien Barthez. The French goalkeeper shouted orders to the wall of players standing 9 meters from the ball. Goal! Silence from the stunned spectators: the ball went around the wall before entering the goal! The TV commentator couldn't believe his eyes:

> *Look: it looks like it's going to go out and, with a bend, it touches the inside of the post . . . A fantastic goal by Roberto Carlos, an incredible bend given to that ball!*

After the match, a player said:

> *We told ourselves that it was the sort of thing you just had to admire and applaud. There are goals that go down in history, and this is one of them.*

It was the effect of the ball's spin that created the lift we have described. Here is what Roberto Carlos himself said:

I always strike my free kicks on the ball's valve, where it is the hardest, to have more power. I've always kicked from the lower left of the ball to the upper right to give it a curved trajectory.

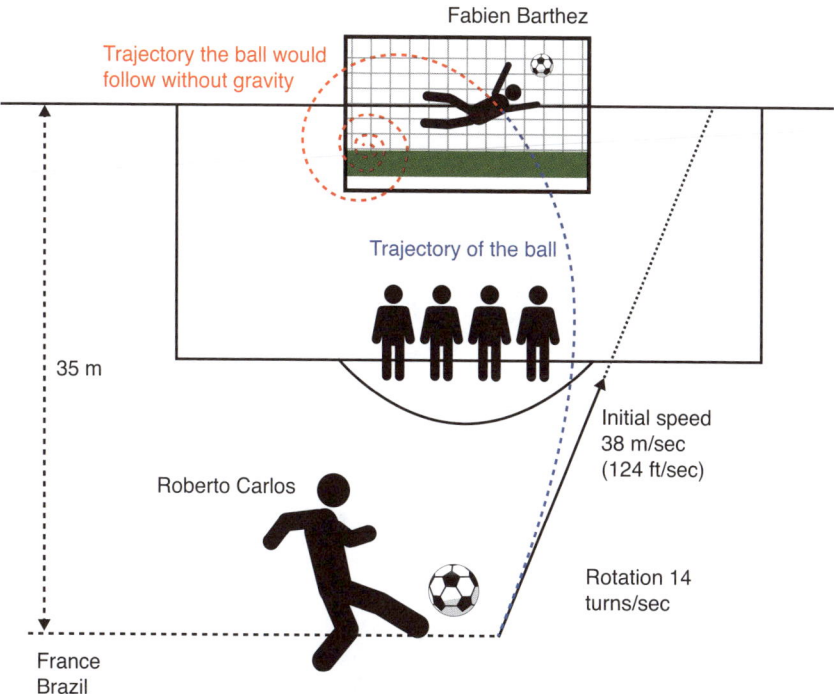

Fabien Barthez

Trajectory the ball would follow without gravity

Trajectory of the ball

35 m

Roberto Carlos

Initial speed 38 m/sec (124 ft/sec)

Rotation 14 turns/sec

France
Brazil

7 Paper, Developables, and Geometry Theorems

Back to geometry!

The old "cubic" balls from before 1930 had to be inflated a lot to have a really round shape. Of course, a cube isn't a ball if its faces aren't deformed.

But the Brazilian ball *Brazuca* almost doesn't need to be inflated. Its faces curve naturally. Once assembled, they are no longer flat.

Designers understood that and tried to improve it even more.

To conclude this little book, let's examine a few theorems in geometry.

AN EXISTENCE THEOREM

To be honest, so far we've managed to avoid talking about a major difficulty. We have drawn many patterns and have suggested that by correctly gluing sides together, as indicated on a pattern, we could in fact construct a polyhedron. That was perhaps clear for a cube, with its six square faces, but it is much less so, for example, for this pattern.

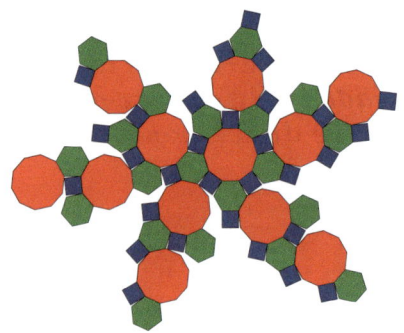

And yet, it works.

Imagine any number of convex polygons spread out on a table. Convexity means that the polygons don't have reflex angles: all the interior angles are less than 180 degrees. The polygon on the left is convex and the one on the right is not.

Let's assume that we can group the sides of these polygons into pairs that have the same length, and we want to glue these pairs to construct a polyhedron. Is this possible?

First, around each of the vertices of the polyhedron we wish to construct, the faces must follow each other cyclically.

If we want a convex polyhedron, the sum of the angles of the faces that meet at each vertex must be less than 360 degrees, or a complete turn.

The theorem, ultimately rather difficult to demonstrate (but which many people consider to be more or less obvious), is the converse.

Every pattern such that the sum of the angles at each vertex is less than 360 degrees can be made into a convex polyhedron in three-dimensional space.

A RIGIDITY THEOREM

Let's make a cardboard pyramid. To do this, we'll begin by cutting a sheet of cardboard as indicated following the SABCDE pattern below, then folding along the dotted lines and, finally, gluing the sides AS and ES. The result is a type of cone whose vertex is the point S and whose edge is a quadrilateral ABCD.

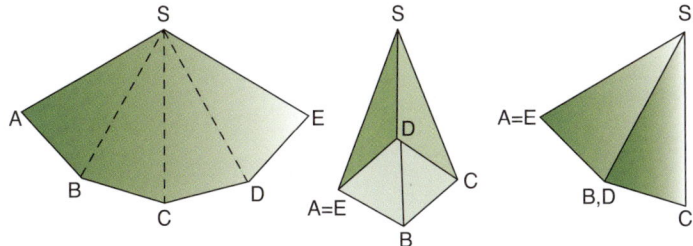

This object is flexible. If you hold it in your hand, the quadrilateral ABCD can be reshaped to be more or less open; the construction isn't very solid. To complete the pyramid, you also have to cut out a cardboard square and glue it onto the quadrilateral to form the base. After this operation, the pyramid becomes solid and rigid. If you put it on a table, it doesn't collapse. If you pick it up and try to deform it (carefully), you can't, unless you deform the cardboard faces. Similarly, a cardboard cube is rigid, as we all know. How about any other polyhedron, one that perhaps has thousands of faces? Are the Amazon Spheres in Seattle rigid? That last question suggests that the subject of rigidity and flexibility is perhaps not merely theoretical.

The subject of the rigidity of these types of objects is very old. Euclid was probably aware of it. The great eighteenth-century French mathematician Adrien-Marie Legendre was interested in the subject and spoke to his colleague Joseph-Louis Lagrange about it; the latter in turn suggested to the young Augustin-Louis Cauchy in 1813 that he study the question. This tip would lead to the first notable findings of Baron Cauchy, who would subsequently become one of the greatest mathematicians of his century. Cauchy was interested in convex polyhedrons, that is, polyhedrons that do not have edges that bend inward. For example, the pyramid that we constructed, or the football, are convex, whereas the following object discovered by Kepler is not.

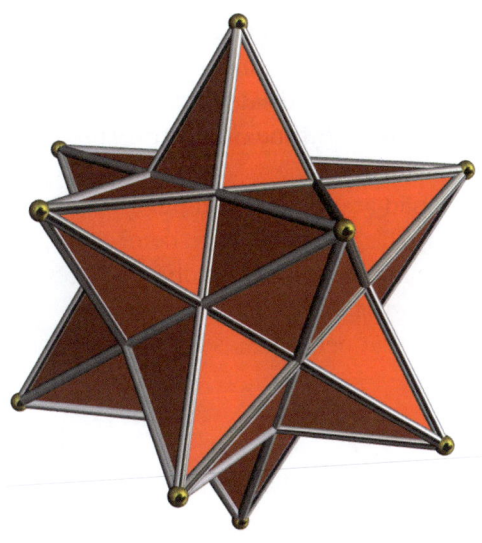

The theorem Cauchy came up with is this:

Every convex polyhedron is rigid.

This means that if we construct a convex polyhedron with unbendable polygons (out of metal, for example) adjusted by hinges along their edges, the global geometry of the whole prevents the joints from moving. The cone we constructed is flexible, but that doesn't invalidate the theorem. It lacks a face, and it is the last face that rigidifies the pyramid. Working in mathematics means demonstrating what one asserts, and Cauchy's demonstration is superb (even if some people subsequently pointed out that it was incomplete). Unfortunately, now is not the time or place to fully examine the proof, but I would like to extract a "lemma" from it, that is, a step in the demonstration.

Let's place a few metal rods assembled end to end on the ground. At each of the angles of this polygonal chain, move the two corresponding rods to decrease the angle. Then the two ends of the chain become closer. Does that seem obvious? Try to demonstrate it.

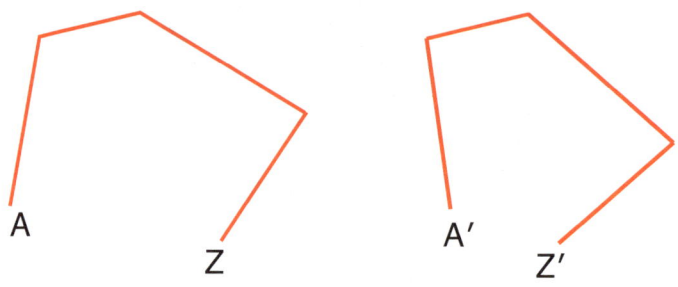

In fact, Cauchy's theorem is more precise. If we have a pattern made up of convex polygons and if the sum of the angles meeting at a vertex is less than a complete turn, then we can construct a convex polyhedron. Well, Cauchy's rigidity theorem says a great deal more: regardless of the way in which one proceeds to glue the faces together, at the end of the construction one always obtains the same polyhedron!

For a long time, many mathematicians wondered if non-convex polyhedrons were equally rigid. Can we find a proof of rigidity that doesn't use the hypothesis of convexity? Mathematicians love statements in which all hypotheses are useful for reaching a conclusion. They had to wait more than one hundred sixty years to have the answer in this particular case. In 1977, mathematician Robert Connelly surprised everyone. He constructed a (rather complicated) polyhedron that was flexible, though not convex (so as not to thwart Cauchy). Since then, his construction has been somewhat simplified, in particular by Klaus Steffen.

Here is the pattern that will enable you to construct Steffen's flexihedron. The sides have five lengths: 5, 10, 11, 12, and 17 (centimeters, for example). Cut it out, and fold along the lines. The solid lines are the edges that bend outward, and the dotted lines correspond to the edges that bend inward. Glue the free edges in the obvious way. You will obtain a sort of bird beak and you will see that it is indeed flexible (a bit).

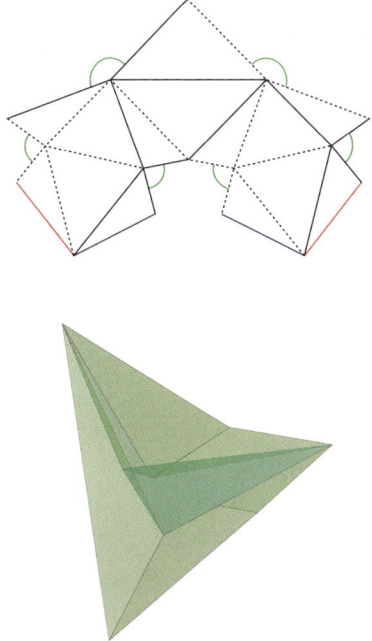

At the time, mathematicians were enchanted by this new object. A metal model was constructed and placed in the tea room of the Institut des Hautes Études Scientifiques in

Bures-sur-Yvette, near Paris, and people could have fun moving this thing. (Actually, it was not very attractive and it squeaked a bit.) The story goes that Dennis Sullivan came up with the idea of blowing cigarette smoke inside Connelly's flexihedron, and he noted that when the object moved, no smoke came out. He then intuited that when the flexihedron is deformed, its volume doesn't vary. Is that story true? Whatever the case, Connelly and Sullivan conjectured that when a polyhedron is deformed, its volume remains constant. It isn't difficult to verify this in the specific example of Connelly's flexihedron or in that of Steffen (at the cost of complicated calculations, which aren't very interesting). However, the conjecture in question refers to all polyhedrons, including those that have never been physically constructed! They called this the "bellows conjecture," because a fireplace bellows ejects air when it is pressed (that is, after all, its function); in other words, its volume diminishes. Of course, an actual bellows is different from the objects Connolly and Sullivan had in mind: it is made of leather and its surfaces are constantly deformed, unlike polyhedrons with rigid faces.

In 1997, Connelly and two other mathematicians, Idzhad Sabitov and Alfred Walz, finally succeeded in proving the conjecture. Their published demonstration is impressive and illustrates once again the interactions that occur between all branches of mathematics. In this eminently geometrical problem, the authors used very sophisticated methods of modern abstract algebra. It didn't involve a demonstration that Cauchy "could have found," for the mathematical techniques of his time were insufficient. I would like to recall a formula that used to be taught in high school.

If the lengths of the sides of a triangle are a, b, and c, one can easily calculate the area of the triangle. To do this, we first calculate the semiperimeter $s = (a+b+c)/2$ and then we obtain the area by extracting the square root of $s(s-a)(s-b)(s-c)$. This nice formula is named after the first-century Greek mathematician Heron and comes to us from the darkness of time. Can we calculate the volume of a polyhedron in a similar way if we know the lengths of its edges? Our three contemporary mathematicians have shown that we can. They start from a polyhedron constructed from a pattern made of a certain number of triangles and they call the lengths of the sides of these triangles (possibly very numerous) l_1, l_2, l_3, etc. They then find that the volume V of the polyhedron must satisfy an equation of the nth degree, that is, an equation that involves the nth power of V, such as

$$a_0 + a_1 V + a_2 V^2 + \cdots + a_n V^n = 0.$$

The number n depends on the pattern used and the coefficients of the equation (a_0, a_1, etc.) depend explicitly on the lengths of the sides l_1, l_2, l_3, etc. In other words, if we know the pattern and the lengths of the sides, we know the equation. If we remember that an equation in general has a solution when it is of the first degree, two solutions when it is of the second degree, we might guess that an equation of degree n has at most n solutions.

In conclusion, if we know the pattern and the lengths, we do not necessarily know the volume, but at least we know that the volume can only assume a finite number of values. When the flexihedron is deformed, its volume thus cannot vary continuously (otherwise, the volume would assume an infinite number of successive values); the volume is stuck, and the bellows conjecture is proven.

ORIGAMI

In Japanese, *ori* and *kami* mean "folding" and "paper." Origami is the art of folding sheets of paper to create beautiful shapes. There are thousands of origami models. Some are very simple, like the paper hats that children enjoy making.

Other origami figures are more complicated and require more experience, like this bird.

Is this geometry or is it art? A bit of both, but it's mainly pleasure!

You can find thousands of origami models on the internet. Animals, airplanes, stars, boxes, and many more. And, of course, there are also patterns to make a paper *Telstar*.

CURVED ORIGAMI

Usually, when you create an origami, you fold sheets of paper, but you don't curve the surfaces which remain planar like the faces of a cube, for example.

Can you make a perfectly round ball with paper? Certainly not. Take a sheet of paper and try to stick it onto a ball, a football, for example. You'll see that the paper crumples and maybe even tears.

Of course, our *Telstar* is essentially a ball, but not entirely. It had to be inflated for the faces to be rounded, thereby deforming them a bit, which isn't practical from a manufacturing point of view. Perhaps you noticed that the *Brazuca* ball has only six faces and that when they are glued together, its shape is already almost that of a ball, without needing to be inflated. Why is that?

Take a sheet of paper in your hands and bend it gently (without tearing it). You'll get a shape like this:

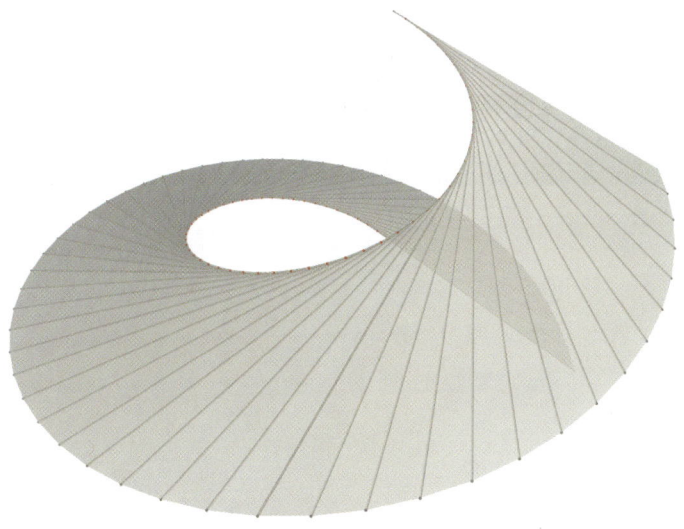

Mathematicians say that this surface is *developable*, meaning that it can be "developed," that is, placed on a table, for example, without tearing or crumpling. The surface above is developable, but the surface of a sphere is not. Cylinders and cones are also developable surfaces.

So the idea is to make origami shapes whose surfaces are not flat but developable. After all, we're using paper, which can be deformed. These are called curved origami. Some are magnificent. Here is an example in the shape of a flower.

A THEOREM BY ALEXANDROV AND POGORELOV

Mathematicians have wondered if it is possible to make curved origami that would be almost ball shaped.

Let's begin with a simple example. A region in a plane is convex if the segment that joins any two of its points is entirely contained in the region. In the figure below, the drawing on the left is convex, and the one on the right is not.

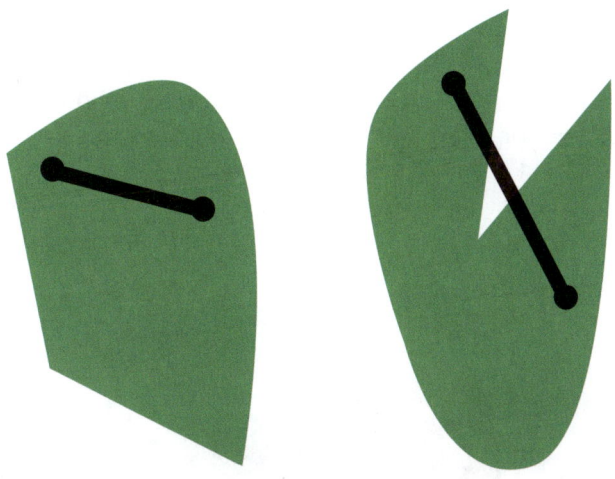

Now, consider two convex regions within the plane whose borders are the same length. Cut these shapes out of paper, choose a point on the edge of each of them, and glue those points together.

Then, with adhesive tape or glue, continue to stick the edges of the two shapes together. In the 1970s, it was shown that you can continue gluing all the way around, that is, that folds never appear at the seam, preventing you from continuing. You will therefore create an object in space that turns out to be convex. This is not at all obvious! You might think that you couldn't glue the two shapes together without tearing them here or there. And there is no obvious reason why the constructed object should be convex.

The object you have constructed is made up of two developable surfaces glued together at their edges.

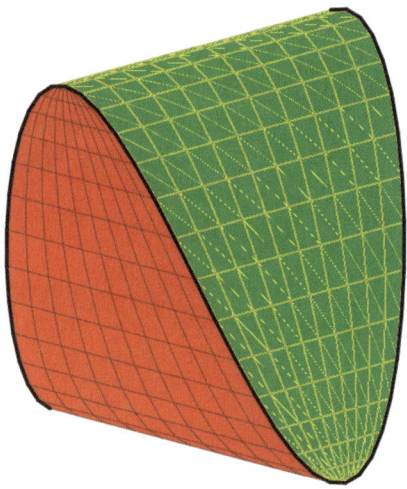

Instead of starting with two convex shapes in the plane, you can, for example, start with six, which you can think of as the "squared faces" of a cube. On the edges of each of these domains, pick four points that you think of as the vertices of a "square." Assume that the four "corners" that you have chosen are in fact corners, that is, that the domains display angles at these vertices.

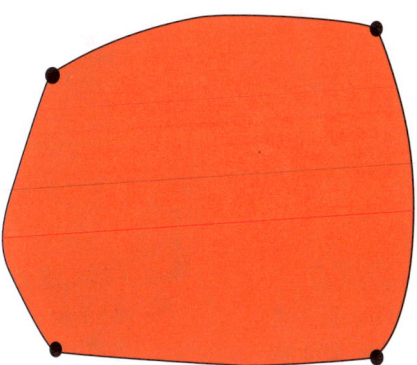

Finally, assume that the lengths of all the "curved edges" are the same. Now take your adhesive tape and stick them all together, as if to make a cube.

We need another hypothesis: each vertex of a cube is common to three faces. The sum of the three corresponding angles must be less than or equal to 360 degrees.

The Alexandrov–Pogorelov Theorem guarantees that, under these hypotheses, this construction works! You'll create a sort of cube, whose edges are curved and whose six faces are developable and not necessarily flat.

In fact, the theorem is even more general and doesn't even presuppose that the faces used are convex. The important condition is that when you glue two faces together on an edge, the sum of the two curvatures of the two faces at each of the points of contact must be positive or zero. In less precise terms, you have to glue a concavity to an equal or stronger convexity.

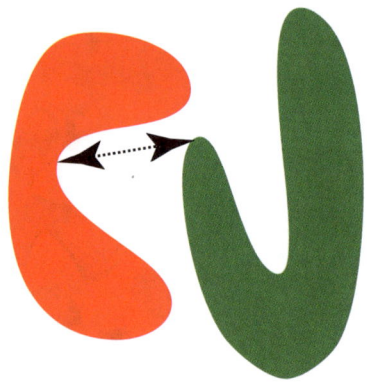

Here is the pattern for the *Al Rihla* ball:

We can see that the triangles and the kites have curved edges. Once these flat panels are glued together, they assume the shape of developable surfaces and produce an almost spherical icosidodecahedron.

The *Brazuca* is created by applying the Alexandrov–Pogorelov Theorem (perhaps unbeknownst to the designers). Here are the six "squared" faces that are glued together to create a "spherical cube."

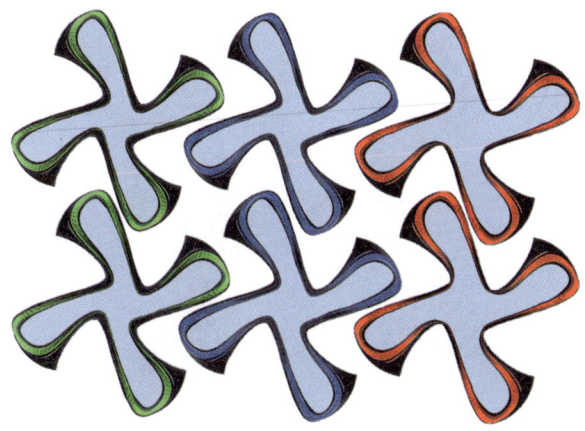

Note that each panel indeed has four "corners" that are seen in black. The angle of each of these corners is 120 degrees.

Amazingly, at the same time the Adidas designers were coming up with *Brazuca*, the great mathematician William Thurston was reaching the same conclusion by another route.

Like many geometers of the past, Thurston was interested in the way in which clothing was cut to cover the surface of the human body. In one of his last articles, "Playing with Surfaces"–a collaboration with Kelly Delp–he explains his ideas and his failures. As a good theoretical mathematician, he assumed that the surface to be dressed was perfectly spherical.

He started with an octahedron and sought the shape to give to the triangular faces so that the resulting object would be as spherical as possible.

Here are the eight "triangles" he proposed, again, almost at the same time the designers were coming up with *Brazuca*. Cut eight shapes like this out of metallic paper.

Then, glue them together as for an octahedron with eight triangles. Here's the result:

It's almost perfect. It's almost round. In the end, all of that is very close to *Brazuca*, right?

William Thurston wasn't content with covering balls with paper. He collaborated with leading designers on a Paris fashion show in 2010. A newspaper's headline: "When Fashion Encounters Mathematics."

To conclude, here is a nice, curved *Telstar*. The hexagons and the pentagons have six and five arms. For the next World Cup, perhaps?

Beautiful Footballs

In this book, we have talked more about geometry than football tactics. Is it useful or interesting to examine the geometry of footballs? What in fact constitutes an interesting mathematical problem? That is a difficult question upon which mathematicians have been reflecting for a long time.

Here are a few elements of a response, a few indications of "quality."

The first criterion is how old something is. Mathematicians are very sensitive to tradition, to problems that have been stated for a long time and that several generations of mathematicians have tried to solve. We have seen that World Cup balls have been inspired by the work of Plato, Leonardo da Vinci, and Kepler, going back well before the invention of football. Their motivations had nothing to do with the sport. They were seeking the symmetries of the universe that would enable them to understand the world. They saw in the sphere a symbol of perfection. For example, Copernicus started his great book of 1543, in which he rebuilt our conception of space in this way:

> *In the beginning we should remark that the world is globe-shaped; whether because this figure is the most perfect of all, . . . or because everything in the world tends to be delimited by this form, as is apparent in the case of drops of water and other liquid bodies, when they become delimited of themselves. And so no one would hesitate to say that this form belongs to the heavenly bodies.*

The question of whether a good problem must have useful applications in practice is more difficult. Mathematicians respond to it in various ways. Undoubtedly, "practical" questions, arising for example from physics, often serve as motivation for mathematics. Sometimes, it's a matter of solving a very concrete problem such as designing a football. But the connection is often less clear. The mathematician then makes use of the concrete problem only as a source of inspiration for other more general questions, and the effective solution of the initial problem is no longer the true goal.

Beyond tradition and utility, a good geometry problem must involve attractive objects. I hope I have convinced the reader that footballs are beautiful!

Nicolaus Copernicus (1473–1543)

Credits

A

Adidas Al Rihla, football created for the 2022 FIFA® World Cup. Adidas website. _ **72 Top Left, 100**

Adidas Beau Jeu, one of the official match balls of the UEFA Euro 2016 tournament, Wikimedia Commons/MÆBOE. _ **65 Bottom**

Adidas Brazuca, the official match ball of the 2014 FIFA® World Cup, Wikimedia Commons/Nicola. _ **65 Top**

Adidas Jabulani, the official match ball for the 2010 FIFA® World Cup, Wikimedia Commons/warrenski. _ **62 Bottom**

Adidas Teamgeist, official match ball for the 2006 FIFA® World Cup in Germany, Armellino Raffaele (CC BY 4.0). _ **60**

Adidas Telstar, official match ball for the 1970 FIFA® World Cup in Mexico, FIFA® website. _ **46, 48 Top Left**

Adidas Tricolore, official match ball of the 1998 FIFA® World Cup in France, Wikimedia Commons/2010 Word Cup–Shine 2010 from Johannesburg, South Africa. _ **48 Bottom Left**

Andreev, Nikolai and Étienne Ghys, Dodecahedron. _ **34-35**

Atrops235, icosidodecahedron, en.wikipedia. _ **73 Bottom**

C

Champetier, Christophe, Smoke. _ **98 Top**

Champions League footballs, 20381303/24 images/Pixabay. _ **16 Bottom**

Chapman-Bell, Philip, curved origami, flick/photos/oschene/. _ **118**

Copernicus, Nicolaus, *On the Revolutions of Heavenly Spheres*, trans. Charles Glenn Wallis (Prometheus Books, 1995), 9. _ **118**

G

Germydan, golf ball, sketchfab.com. _ **101 Top**

Ghys, Étienne. _ **13, 16 Top, 17 Top, 29 Bottom, 30 Top, 37 Middle, 39 Middle and Bottom, 41, 58 Bottom, 62 Top, 64 Bottom, 66 Top, 67 Bottom, 69, 72 Top Right and Top Middle, 73 Top, 74 Bottom, 75, 77, 82, 83 Bottom, 86 Right, 88, 89, 90, 92, 94, 97, 105, 108 Bottom, 109, 112, 113 Bottom, 117, 119, 120, 121, 122, 123**

Ghys, Étienne, modified from academic.com (https://fr-academic.com/dic.nsf/frwiki/1091). _ **49 Left**

Ghys, Étienne, modified from Inkscape. _ **59 Bottom**

Ghys, Étienne, modified from Sketchfab, "Adidas Teamgeist Ball (Germany 2006 Match Ball) 3D Model," Created by Armellino Raffaele (CC BY 4.0). _ **59 Top**

Ghys, Étienne, modified from Wikimedia Commons:
Bauer, Andreas, origami, origami-kunst.de. _ **116 Top**

Bernard de Go Mars, flows over the sphere. _ **96, 98 Bottom**

Cmglee, truncated icosahedron. _ **45**

Fetti, Domenico, Archimedes. _ **52**

Kgbo, Copernicus. _ **128**

Köhler, August, Johannes Kepler.- **33 Bottom**

Leonardo da Vinci, *Divina proportione* by Luca Pacioli (1509). _ **40, 49 Top Right**

Rdurkacz, Magnus effect. _ **103**

Sailko, Portrait of Galileo by Sustermans (photo B. Mollier). _ **87**

Watchduck/Tilman Piesk, Archimedean solids. _ **53-56, 58 Top, 61 Bottom, 108 Top**

Greatpatton, Football pitch, Wikipédia.fr. _ **81**

H

© Hyber, Fabrice, *Square football* (with his kind permission). _ **19**

Franz, Opening ceremony for the 2006 World Cup. _ **57**

Fropuff Mysid, Golden rectangles. _ **37 Bottom**

Gustave Eiffel, "Crisis of drag" from *New Research on Air Resistance and Aviation Made at the Auteuil Laboratory* (Dunod and Pinat, 1914). _ **95**

Haeckel, *Kunstformen der Natur*, 1904. _ **36**

HeliumPlasma, Polyhedron patterns. _ **22 Bottom, 26 Bottom, 27 Bottom, 28 Bottom, 30 Bottom**

Here, Poster of Henri Desmé, 1938 World Cup. _ **66 Bottom Right**

Hiku2, Dodecahedron calendar. _ **36**

Logo 1970 FIFA® World Cup in Mexico. **48 Top Right**

Logo 1998 FIFA® World Cup in France. _ **48 Bottom Right**

Mainstone, John, University of Queensland. File created by Amada44 / Wikimedia Commons (CC BY-SA 3.0), Tar drop. – **93**

MDBR, Footballs 1938 and 1962. _ **66 Bottom Left, 67 Top Left**

Nonmacher, R. A., Cube patterns. _ **22**

Phil Clement, personal work, Logo UEFA Champions League®. _ **17 Bottom**

Platonic dice. _ **39 Top**

Rotenberg, truncated icosahedron. **44**

Sim3331621, Flexidron. _ **113 Top**

Splettstoesser, Thomas, Adenovirus. _ **37 Top**

Tomruen, Uniform tilings, English Wikipedia. _ **23, 27, 29 Top, 31, 74 Top**